IGCSE
Mathematics

module 4

University of Cambridge Local Examinations Syndicate

Reviewed by **John Pitts**, Principal Examiner and Moderator for HIGCSE Mathematics

Edited by **Carin Abramovitz**

PUBLISHED BY THE PRESS SYNDICATE OF THE UNIVERSITY OF CAMBRIDGE
The Pitt Building, Trumpington Street, Cambridge CB2 1RP, United Kingdom

CAMBRIDGE UNIVERSITY PRESS
The Edinburgh Building, Cambridge CB2 2RU, UK http://www.cup.cam.ac.uk
40 West 20th Street, New York, NY 10011-4211, USA http://www.cup.org
10 Stamford Road, Oakleigh, Melbourne 3166, Australia
Dock House, Victoria and Alfred Waterfront, Cape Town 8001, South Africa

© University of Cambridge Local Examinations Syndicate 1998

This book is in copyright. Subject to statutory exception
and to the provisions of relevant collective licensing agreements,
no reproduction of any part may take place without
the written permission of Cambridge University Press.

First published 1998
Fourth printing 2002

Printed by Creda Communications, Cape Town

Typeface New Century Schoolbook 11.5/14 pt

A catalogue record for this book is available from the British Library

ISBN 0 521 62516 5 paperback

Acknowledgements
We would like to acknowledge the contribution made to these materials by the writers and editors of the Namibian College of Open Learning (NAMCOL).

Illustrations by André Plant.

Contents

Introduction iv

Unit 1   **Geometrical Terms and Facts**   1
  A   Angles   1
  B   Triangles, quadrilaterals and other polygons   17
  C   Symmetry   30
  D   Circles   39

Unit 2   **Practical Geometry**   47
  A   Using geometrical instruments   47
  B   Scale drawings   62
  C   Geometrical constructions   72
  D   Loci   74

Unit 3   **More Terms and Facts**   87
  A   Solid figures   87
  B   Congruent figures and similar figures   99
  C   More results about circles   110

Solutions   123

Index   141

Introduction

Welcome to Module 4 of IGCSE Mathematics! This is the **fourth module** in a course of six modules designed to help you prepare for the International General Certificate of Secondary Education (IGCSE) Mathematics examinations. Before starting this module, you should have completed Module 3. If you are studying through a distance-education college, you should also have completed the **end-of-module assignment** for Module 3. The diagram below shows how this module fits into the IGCSE Mathematics course as a whole.

Module 1	Module 2	Module 3	Module 4	Module 5	Module 6
Assignment 1	Assignment 2	Assignment 3	Assignment 4	Assignment 5	Assignment 6

Like the previous module, this module should help you develop your mathematical knowledge and skills in particular areas. If you need help while you are studying this module, contact a **tutor** at your college or school. If you need more information on writing the examination, planning your studies, or how to use the different features of the modules, refer back to the **Introduction** at the beginning of Module 1.

Some study tips for Maths

- As you work through the course, it is very important that you use a **pen or pencil and exercise book**, and *work through the examples yourself* in your exercise book as you go along. Maths is not just about reading, but also about doing and understanding!
- Do feel free to write in pencil in this book – fill in steps that are left out and make your own notes in the margin.
- *Don't expect to understand everything the first time you read it.* If you come across something difficult, it may help if you read on – but make sure you come back later and go over it again until you understand it.
- You will need a **calculator** for doing mathematical calculations and a **dictionary** may be useful for looking up unfamiliar words.

Remember

- In the examination you will be required to give decimal approximations correct to **three significant figures** (unless otherwise indicated), e.g. 14.2 or 1 420 000 or 0.00142.
- Angles should be given to **one decimal place**, e.g. 43.5°. Try to get into the habit of answering in this way when you do the exercises.

The **table** below may be useful for you to keep track of where you are in your studies. Tick each block as you complete the work. Try to fit in study time whenever you can – if you have half an hour free in the evening, spend that time studying. Every half hour counts! You can study a **section**, and then have a break before going on to the next section. If you find your concentration slipping, have a break and start again when your mind is fresh. Try to plan regular times in your week for study, and try to find a quiet place with a desk and a good light to work by. Good luck with this module!

IGCSE MATHEMATICS MODULE 4

Unit no.	Unit title	Unit studied	'Check your progress' completed	Revised for exam
1	Geometrical Terms and Facts			
2	Practical Geometry			
3	More Terms and Facts			

Unit 1
Geometrical Terms and Facts

In this unit you begin your study of geometry. I'll start with the basics, showing you how to draw and measure angles and recognise different shapes. You'll also learn about symmetry and I'll end off the module with a section on the geometry of the circle.

This unit is divided into four sections:

Section	Title	Time
A	Angles	4 hours
B	Triangles, quadrilaterals and other polygons	3 hours
C	Symmetry	2 hours
D	Circles	2 hours

By the end of this unit, you should be able to:
- name and describe angles, triangles, quadrilaterals and polygons
- draw and measure angles
- calculate angles in diagrams containing intersecting lines, parallel lines, triangles, quadrilaterals and/or polygons
- recognise and describe symmetry of plane figures
- use terms associated with circles
- calculate angles in diagrams containing circles.

A Angles

Measurement of angles

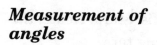

This is an angle. It is formed by two line segments, AB and BC, placed so that they have one end (B) in common.

AB and BC are called the **arms** of the angle, and the common end (B) is called the **vertex** of the angle.

The size of the angle is the amount by which one of the line segments needs to be rotated about the common end so that it lies on top of the other line segment.

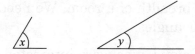

Notice that the size of an angle does not depend on the lengths of the arms. In these diagrams, angle x is bigger than angle y.

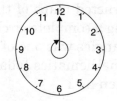

In one hour, the minute hand of a clock makes one complete revolution. This is called **one turn**. Other angles can be expressed as fractions of a turn.

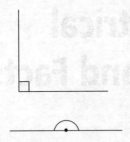

A quarter turn ($\frac{1}{4}$ turn) is called a **right angle**. A right angle is usually indicated by a square symbol, as shown in the diagram.

A half turn ($\frac{1}{2}$ turn) is called a **straight angle** because the two arms of the angle make one straight line.

Example 1

Through what angle does the minute hand of a clock rotate in 25 minutes?

Solution

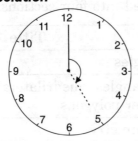

In 60 minutes, the minute hand makes one complete revolution, that is 1 turn.

So, in 1 minute it rotates $\frac{1}{60}$ turn.

In 25 minutes it rotates $\frac{25}{60}$ turn, that is $\frac{5}{12}$ turn.

Example 2

What is the angle between the hands of a clock at 2 o'clock?

Solution

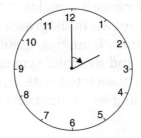 

Notice that there are *two* angles formed by the hands of the clock.

In 2 hours the hour hand rotates $\frac{2}{12}$ turn, that is $\frac{1}{6}$ turn. Hence, the smaller angle between the hands at 2 o'clock = $\frac{1}{6}$ turn.

The larger angle and the smaller angle add up to 1 complete turn. Hence, the larger angle between the hands at 2 o'clock = $\frac{5}{6}$ turn.

Most of the angles in the work we do are less than a complete turn and so their sizes are fractions of a turn. This is inconvenient and arises because a unit of one turn is so large. The situation is similar to using a unit of one kilometre to measure distances such as the length and breadth of a room. We need to find a unit smaller than one turn for angles.

The Babylonians, who were interested in astronomy as well as mathematics, constructed maps of the heavens and measured directions by dividing a complete circle (or turn) into 360 parts. Their unit for measuring angles was one of these parts and it is the one we use in elementary mathematics today. It is known as one **degree**, and is usually written as 1°.

There are 360 degrees in one complete turn. We write this as:

$$1 \text{ turn} = 360°$$

> There are other units for measuring angles known as 1 **grad** (sometimes called 1 gradian) and 1 **radian**. Your calculator can be programmed to work in degrees or grads or radians but, for the time being, you should concentrate on degrees and turns.

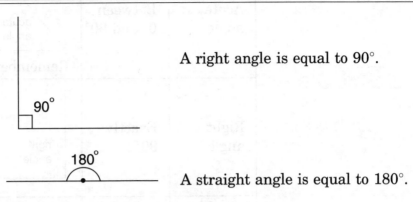

A right angle is equal to 90°.

A straight angle is equal to 180°.

Example 1

Find, in degrees, the size of an angle equal to $\frac{3}{8}$ turn.

Solution

$$\begin{aligned}
1 \text{ turn} &= 360° \\
\tfrac{3}{8} \text{ turn} &= \tfrac{3}{8} \times 360° \\
&= 3 \times 45° \\
&= 135°
\end{aligned}$$

Example 2

Find, as a fraction of one turn, the size of an angle equal to 200°.

Solution

$$\begin{aligned}
360° &= 1 \text{ turn} \\
1° &= \tfrac{1}{360} \text{ turn} \\
200° &= \tfrac{200}{360} \text{ turn} \\
&= \tfrac{20}{36} \text{ turn} \\
&= \tfrac{5}{9} \text{ turn}
\end{aligned}$$

Example 3

Find, as a fraction of one turn, the size of an angle equal to 270°.

Solution

$$\begin{aligned}
90° &= 1 \text{ right angle} = \tfrac{1}{4} \text{ turn} \\
270° &= 3 \times 90° \\
&= \tfrac{3}{4} \text{ turn}
\end{aligned}$$

Types of angles

Type	Size	Drawings
Acute angle	Between 0° and 90°	Remember: 'acute' means 'sharp'
Right angle	Exactly 90°	
Obtuse angle	Between 90° and 180°	
Straight angle	Exactly 180°	
Reflex angle	Between 180° and 360°	

Naming angles

In this diagram, the marked angle can be named as angle A or $\hat{A}$ or $\angle A$.

Sometimes an angle is named by using a single small letter, such as x in this diagram.

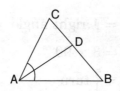

In this more complicated diagram, we cannot use *angle A* to name the marked angle because there is more than one angle at the point A.
In this case, we use a three-letter notation. The marked angle is named as
    angle BAC   or $B\hat{A}C$ or $\angle$ BAC.
Notice that the vertex of the angle is given by the letter in the middle.

Example 1

Estimate the size, in degrees, of the following angles.

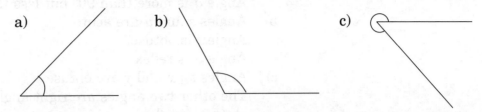

a)   b)   c)

Solution

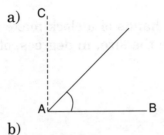

a) Through the vertex A, draw a line AC at right-angles to the arm AB. The given angle is roughly half the right angle, so its size is about one half of 90°.

The angle is about 45°.

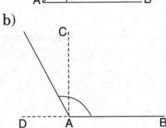

b) Extend the arm BA beyond A to the point D and, through the vertex A, draw a line AC at right-angles to the arm AB.
The given angle is more than the right angle BAC but less than the straight angle BAD. It is roughly $90° + \frac{1}{3}$ of 90°.
The angle is about 120°.

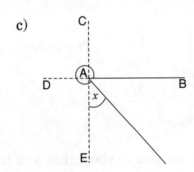

c) Extend the arm BA beyond A to a point D and, through the vertex A, draw lines AC and AE at right-angles to the arm AB.
The given angle is equal to 3 right angles + angle x.
Angle x is roughly half a right angle, so the given angle is roughly $3 \times 90° + 45°$.
The angle is about 315°.

Example 2

State which of the angles marked with a small letter in the following diagrams are acute, which are obtuse and which are reflex.

a) 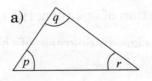   b) 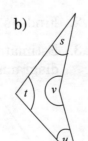   c)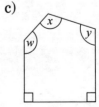

Solution

a) Angles p and r are both less than 90°. They are acute.
   Angle q is more than 90° but less than 180°. It is obtuse.
b) Angles s and u are acute.
   Angle t is obtuse.
   Angle v is reflex.
c) Angles w, x and y are obtuse.
   The other two angles are right angles (so they are not acute or obtuse or reflex).

Example 3

At 5 o'clock the hands of a clock make an obtuse angle and a reflex angle. Calculate the size, in degrees, of each of these angles.

Solution

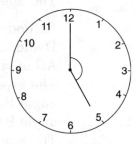

The obtuse angle $= \frac{5}{12}$ of a turn
$= \frac{5}{12} \times 360°$
$= 150°$

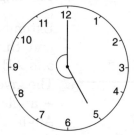

The reflex angle $= \frac{7}{12}$ of a turn
$= \frac{7}{12} \times 360°$
$= 210°$

Now it's time for you to show that you have understood the work so far. Try this exercise.

EXERCISE 1

1. Find, in degrees, the size of an angle equal to $\frac{5}{6}$ turn.

2. Find, as a fraction of one turn, the size of an angle equal to 315°.

3. Estimate the size, in degrees, of the angles marked in these diagrams.

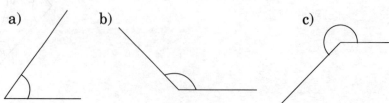

4. State which of the angles marked with a small letter in the following diagrams are acute, which are obtuse and which are reflex.

a) 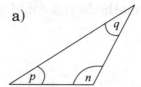   b) 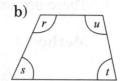   c)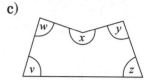

5. At 10 o'clock the hands of a clock make an acute angle and a reflex angle. Calculate the size, in degrees, of each of these angles.

Check your answers at the end of this module.

Measuring angles

To measure angles, you use a geometrical instrument called a **protractor**.

It is marked in degrees. It has two scales, a clockwise scale and an anti-clockwise scale. The clockwise scale is on the outside and starts at 0 on the left, increasing in a clockwise direction. The anti-clockwise scale is on the inside and starts at 0 on the right, increasing in an anti-clockwise direction.

The protractor has a centre point and a zero line, as shown in this diagram. (The zero line is not usually the bottom edge of the protractor.)

To measure an angle less than 180°, follow these steps:

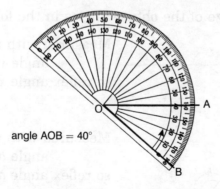

angle AOB = 40°

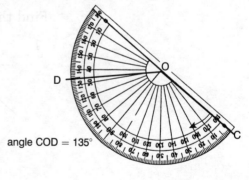

angle COD = 135°

- Estimate the size of the angle – decide whether it is less than a right angle (90°) or more than a right angle.
- Place the protractor on the angle so that its centre point is on top of the vertex of the angle and its zero line is on top of one arm of the angle. (If necessary, extend the arms of the angle so that they reach the outer edge of the protractor.)
- Start from the 0 (zero) mark which is on top of the arm of the angle. Notice whether it is on the clockwise scale or the anti-clockwise scale. Go round its scale until you reach the other arm of the angle. Count the degrees on the scale as you go round: 0, 10, 20, ..., then any extra degrees 1, 2, 3, ...

- Write down the total number of degrees.
- Compare the total number of degrees with your estimate – is your answer reasonable?

There are two methods you could use to measure a reflex angle.

Method 1

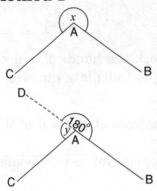

Suppose you want to measure the reflex angle x in this diagram.

Extend the arm BA beyond A to a point D. With your protractor, measure the angle DAC (marked y in this diagram). Then, the reflex angle $x = 180° + y$.

Method 2

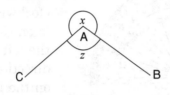

With your protractor, measure the angle BAC (marked z in this diagram). Angle x and angle z together make one complete turn.

$$x + z = 360°$$
$$\text{so } x = 360° - z$$

That is: to find reflex angle x, you measure angle z and take it away from $360°$.

Examples and solutions

Find the size of the obtuse angles in the following:

1.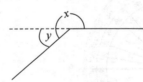

   Measured with a protractor,
   $$\text{angle } y = 32°$$
   so reflex angle $x = 180° + 32°$
   $$= 212°$$ `method 1`

2.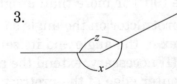

   Measured with a protractor,
   $$\text{angle } b = 126°$$
   so reflex angle $a = 180° + 126°$
   $$= 306°$$ `method 1`

3. Measured with a protractor,
   $$\text{angle } z = 148°$$
   so reflex angle $x = 360° - 148°$
   $$= 212°$$ `method 2`

4.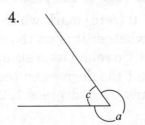

   Measured with a protractor,
   $$\text{angle } c = 54°$$
   so reflex angle $a = 360° - 54°$
   $$= 306°$$ `method 2`

Drawing angles

To draw an angle, you must have a protractor, a ruler and a *sharp* pencil. Suppose you have a straight line segment AB and you want to draw an angle of 125° with its vertex at A and with AB as one of its arms.

Follow these steps:

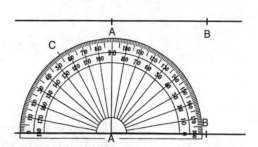

- Put the centre point of your protractor on top of the vertex (A) of the angle, and the zero line of the protractor on top of the arm AB.

- Find the 0 (zero) mark on the protractor which is on top of the arm AB and count round its scale: 0, 10, 20, ... , 110, 120, then the extra degrees 1, 2, 3, 4, 5.

- Put a dot (C) next to the mark 125.

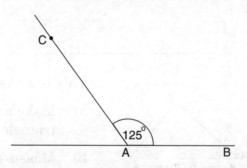

- Remove your protractor and use your ruler to draw a straight line from A and through C.

- Angle BAC is the angle you wanted. Check that it looks about the right size.

Now you should have some practice in measuring and drawing angles. Remember that you should always use a sharp pencil and that your work must be neat. It is essential that you measure and draw angles accurately – errors of more than 1° are unacceptable.

EXERCISE 2

1. Use your protractor to measure each of the angles marked in these diagrams.

   a)   b)   c)

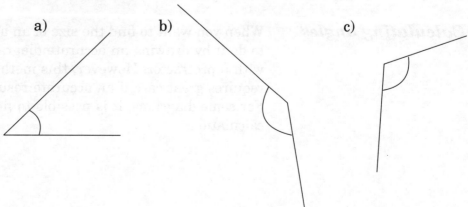

2. Use your protractor to measure the reflex angles marked in these diagrams.

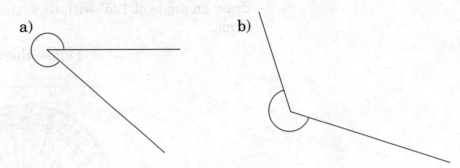

a)

b)

3. Use your protractor to measure each of the angles marked by a small letter in these diagrams. (You will need to extend some of the lines.)

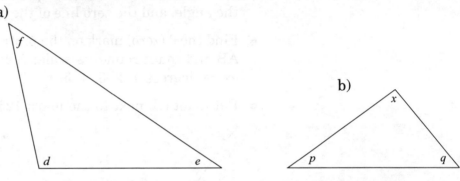

a)

b)

4.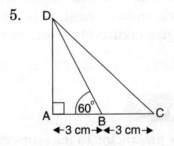

This is a sketch of a triangle XYZ.

a) Make an accurate drawing of the triangle.

b) Measure the size of angle Z of the triangle.

5. Here is another sketch.

a) Make an accurate drawing of the diagram. (Remember that ∟ denotes an angle of 90°.)

b) Measure the size of angle BCD.

c) Measure the length of CD.

Check your answers at the end of this module.

Calculating angles

When you want to find the size of an angle in a sketch, it is possible to do it by drawing an accurate diagram and measuring the angle with a protractor. However, this method is time-consuming and requires great care if an accurate result is to be obtained.

For some diagrams, it is possible to find the size of angles by calculation.

Angles at a point

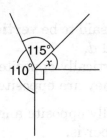

In this sketch, the four angles together make one complete turn, so they must add up to 360°.

$$x + 90° + 110° + 115° = 360°$$
$$x + 315° = 360°$$
$$\text{so } x = 45°$$

This method can be used no matter how many angles meet at a point. We can state the general result:

> Angles at a point add up to 360°.

Angles on one side of a straight line

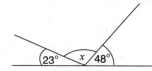

In this sketch, the three angles together make a straight angle, so they must add up to 180°.

$$23° + x + 48° = 180°$$
$$x + 71° = 180°$$
$$\text{so } x = 109°$$

This method can be used no matter how many angles make up the straight angle. The general result is usually stated as:

> Angles on a straight line add up to 180°.

Angles making a right angle

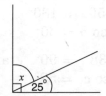

In this sketch, the two angles together make a right angle, so they must add up to 90°.

$$x + 25° = 90°$$
$$\text{so } x = 65°$$

The general result is stated as:

Lines at right-angles to each other are said to be **perpendicular**.

> Angles in a right angle add up to 90°.

Vertically opposite angles

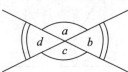

When two straight lines cross one another, they form four angles (a, b, c, d).

You will not be surprised to find that the angles a and c are the same size and that the angles b and d are the same size.

We can show this without doing any measuring, as follows:

a and b are angles on a straight line
so $a + b = 180°$

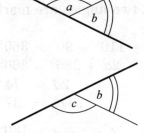

b and c are angles on a straight line
so $b + c = 180°$

It follows that $a + b = b + c$
and hence $a = c$

You should be able to show that $b = d$ in the same way.

Angles a and c are said to be **vertically opposite** to each other, and so are angles b and d.

They are called vertically opposite angles because they have the same vertex and they are opposite to each other.

Notice that vertically opposite angles make an X shape.
The fact to remember is:

> Vertically opposite angles are equal.

Example 1

Calculate the size of each lettered angle in these sketches.

a) 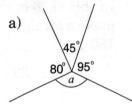 b) 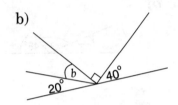 c)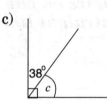

Solution

a) $a + 80° + 45° + 95° = 360°$ (angles at a point)
$a + 220° = 360°$
so $a = 140°$

you must always explain which geometry fact you are using to calculate the angle

b) $20° + b + 90° + 40° = 180°$ (angles on a straight line)
$b + 150° = 180°$
so $b = 30°$

c) $c + 38° = 90°$ (angles in a right angle)
so $c = 52°$

Example 2

Calculate the size of each lettered angle in these sketches.

a) 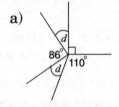 b) 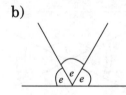 c)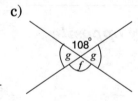

Solution

a) The fact that two angles are marked d indicates that they are equal in size.
$d + 86° + d + 110° + 90° = 360°$ (angles at a point)
$2d + 286° = 360°$
$2d = 74°$
so $d = 37°$

b) $e + e + e = 180°$ (angles on a straight line)
$3e = 60°$
so $e = 60°$

Module 4 Unit 1

c)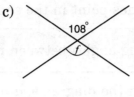

$f = 108°$ (vertically opposite angles)

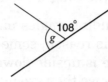

$g + 108° = 180°$ (angles on a straight line)
so $g = 72°$

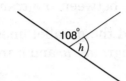

$h + 108° = 180°$ (angles on a straight line)
so $h = 72°$

Example 3

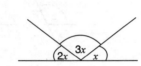

This diagram shows the sizes of three angles on a straight line.

Calculate the value of x.

Solution

$$2x + 3x + x = 180° \text{ (angles on a straight line)}$$
$$6x = 180°$$
$$\text{so } x = 30°$$

Angles associated with parallel lines

Parallel lines are straight lines which are always the same distance apart. The lines are in the same direction as one another. In diagrams, you mark parallel lines with matching arrow heads.

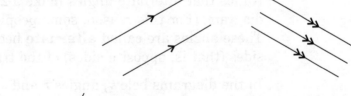

'AB is parallel to CD' is sometimes written as 'AB ∥ CD'.

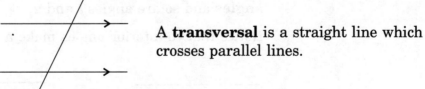

A **transversal** is a straight line which crosses parallel lines.

The transversal and the parallel lines form pairs of **corresponding angles**, as shown in the diagrams below.

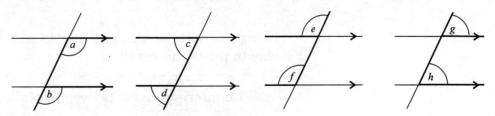

Because parallel lines point in the same direction, it follows that:

> Corresponding angles between parallel lines are equal.

This means that, in the diagrams, $a = b$, $c = d$, $e = f$ and $g = h$.

Notice that corresponding angles make an F shape or a ⅃ shape in the diagram. (For this reason, some people call them 'F angles'.) Sometimes the shape is upside down. These angles are called **corresponding** because they are in corresponding positions (that is, matching positions) between the transversal and the parallel lines.

The transversal and the parallel lines also form pairs of **alternate angles**. In these diagrams, w and x are alternate angles, and so are y and z.

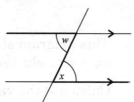
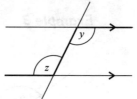

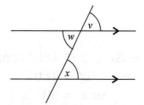

In this diagram,
$w = v$ (vertically opposite angles)
$v = x$ (corresponding angles)
It follows that $w = x$.
Similarly, you can prove that $y = z$.
So, we have the general result:

> Alternate angles between parallel lines are equal.

Notice that alternate angles make a Z shape or a ⟨ shape in the diagram. (For this reason, some people call them 'Z angles'.) These angles are called **alternate** because they are on alternate sides (that is, opposite sides) of the transversal.

In the diagrams below, angles r and s are a pair of **co-interior angles** and so are angles t and u.

Notice that co-interior angles make a ⊏ shape in the diagram.

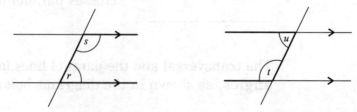

Using alternate angles and angles on a straight line, you should be able to prove the result:

> Co-interior angles between parallel lines add up to 180°.

Module 4 Unit 1

Example 1

Calculate the size of each lettered angle in these sketches.

a)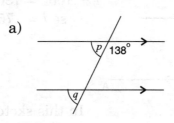
b)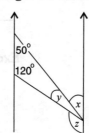

Solution

a)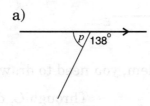

$p + 138° = 180°$ (angles on a straight line)
so $p = 42°$

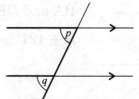

$q = p$ (corresponding angles –
so $q = 42°$ notice the F shape)

b)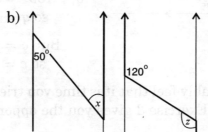

$x = 50°$ (alternate angles – notice the Z shape)
$z = 120°$ (alternate angles)

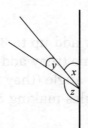

$x + y + z = 180°$ (angles on a straight line)
$50° + y + 120° = 180°$
$y + 170° = 180°$
so $y = 10°$

Example 2

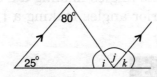

Calculate the size of each lettered angle in this sketch.

Solution

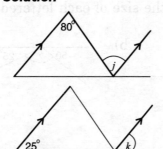

$j = 80°$ (alternate angles)

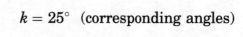

$k = 25°$ (corresponding angles)

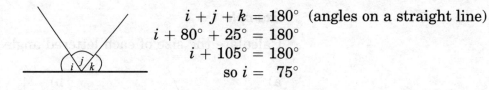

$i + j + k = 180°$ (angles on a straight line)
$i + 80° + 25° = 180°$
$i + 105° = 180°$
so $i = 75°$

Example 3

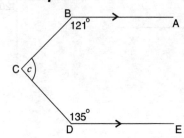

In this sketch, BA is parallel to DE and angle BCD = c.

Calculate the value of c.

Solution

To solve this problem, you need to draw an extra line in the sketch.

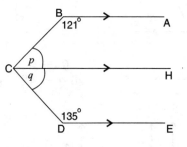

Through C, draw the line CH parallel to BA and DE.

$p + 121° = 180°$ (co-interior angles – notice the [shape)
so $p = 59°$

$q + 135° = 180°$ (co-interior angles)
so $q = 45°$

But $c = p + q$
so $c = 104°$

You probably feel that it is time you tried some angle calculations for yourself. Exercise 3 gives you the opportunity to show what you can do.

Remember to look for:
- angles at a point (they add up to 360°)
- angles on a straight line (they add up to 180°)
- angles making a right angle (they add up to 90°)
- vertically opposite angles making an X shape (they are equal)
- parallel lines and
    (i) corresponding angles making an F shape (they are equal)
    (ii) alternate angles making a Z shape (they are equal)
    (iii) co-interior angles making a [shape (they add up to 180°).

EXERCISE 3

1. Calculate the size of each lettered angle in these sketches.

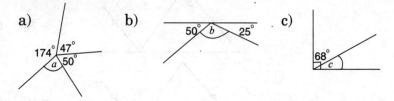

2. Calculate the size of each lettered angle in these sketches.

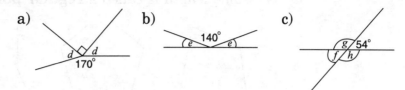

3.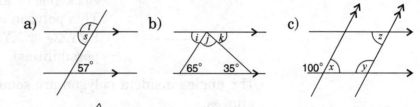

 The diagram shows two intersecting lines.

 Calculate the values of x, p and q.

4. Calculate the size of each lettered angle in these sketches.

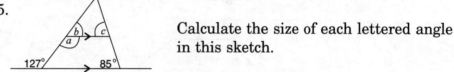

5. Calculate the size of each lettered angle in this sketch.

Check your answers at the end of this module.

B Triangles, quadrilaterals and other polygons

A closed shape whose boundary consists of straight lines is called a **polygon**. The word 'polygon' comes from Greek words meaning 'many angles'.

Some polygons have special names, as shown in the table below. A polygon has the same number of sides as it has angles. The names indicate how many angles or how many sides the polygon has.

Number of sides/angles	Name
3	triangle
4	quadrilateral
5	pentagon
6	hexagon
7	heptagon
8	octagon
9	nonagon
10	decagon

A polygon which has all its angles the same size *and* all its sides the same length is called a **regular polygon**.

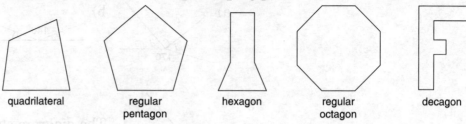

quadrilateral   regular pentagon   hexagon   regular octagon   decagon

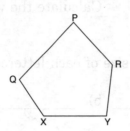

> The plural of **vertex** is **vertices**.

Each corner of the polygon is called a **vertex**.

We refer to a polygon by writing its vertices in the order they occur round the polygon (either clockwise or anti-clockwise).
This polygon could be named PQXYR or PRYXQ or XYRPQ (and there are other possibilities).

The angles inside a polygon are sometimes called the **interior angles**.

If all the interior angles are less than 180°, the polygon is said to be **convex**.
(All *regular* polygons are convex.)

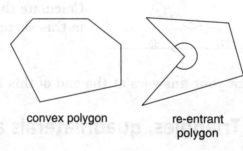

convex polygon   re-entrant polygon

If one or more of the interior angles is reflex, the polygon is said to be **re-entrant**.

If you make a side of a convex polygon longer, the angle it makes with the next side is called an **exterior angle**.

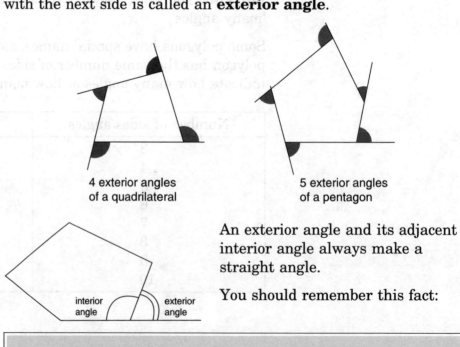

4 exterior angles of a quadrilateral   5 exterior angles of a pentagon

interior angle   exterior angle

An exterior angle and its adjacent interior angle always make a straight angle.

You should remember this fact:

> An exterior angle + adjacent interior angle = 180°.

The sum of the interior angles of a polygon is usually called its **angle sum**. As you will see, it is easy to find the angle sum of a polygon if you know how many sides the polygon has.

Triangles

As its name indicates, a triangle has 3 angles, and so it has 3 sides.

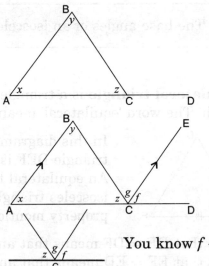

This diagram shows a triangle ABC with the side AC extended to D.

The interior angles of the triangle are x, y and z.

Angle BCD is one of the exterior angles.

A line CE is drawn through C parallel to AB. This divides the exterior angle BCD into two parts, f and g.

You know $f + g + z = 180°$ (angles on a straight line)

$f = x$ (corresponding angles)

$g = y$ (alternate angles)

It follows that $x + y + z = 180°$

In words, this is:

The angle sum of a triangle is 180°.

From the statements above, we can also deduce that
exterior angle BCD $= f + g$
$= x + y$

Notice that x and y are the two interior angles of the triangle, which are opposite to the exterior angle BCD.

We can now state another important result about triangles:

Each exterior angle of a triangle is equal to the sum of the two opposite interior angles.

Isosceles triangle

> In this diagram, a single mark is used to show equal sides, but double or triple marks could be used. This is essential when there is more than one set of equal lines in a diagram.

A triangle with two sides equal in length is called an **isosceles triangle**. The word 'isosceles' comes from a Greek word meaning 'equal legs'.

In the diagram, AB = AC so triangle ABC is isosceles.

Equal sides are indicated by putting a mark on each of them.

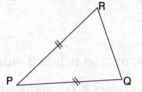

The base angles are opposite the equal sides – they are not necessarily at the base of the triangle.

In this isosceles triangle, angle R = angle Q.

The angles B and C, which are opposite the equal sides, are called the **base angles** of the isosceles triangle. In fact, these base angles are equal in size. (You can see this if you imagine triangle ABC to be cut out and turned over, so that BA and CA exchange places – angle B will fit exactly into the slot which angle C occupied.)

This is an important fact to remember:

> The base angles of an isosceles triangle are equal.

Equilateral triangle

An **equilateral triangle** is a triangle with all its three sides equal in length. The word 'equilateral' means 'equal sides'.

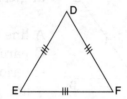

In this diagram, DE = EF = FD and so triangle DEF is equilateral.
An equilateral triangle is a special sort of isosceles triangle and so we can use the property mentioned above.

The fact that DE = DF means that angle E = angle F.
The fact that EF = ED means that angle D = angle F.

It follows that angle D = angle E = angle F.

But, the angle sum of *any* triangle is 180°, so we can deduce that each of the angles D, E, F must be 60°.
This is another important fact for you to remember:

> Each angle of an equilateral triangle is 60°.

You now know that an isosceles triangle is a triangle with two sides equal and that an equilateral triangle is a triangle with all three sides equal. You may also meet the term **scalene triangle**. This type of triangle has all its three sides of different lengths.

Example 1

Calculate the size of each lettered angle in these sketches.

a)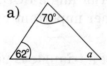

b)

c)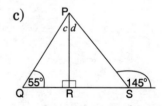

Solution

a) $a + 62° + 70° = 180°$ (angle sum of triangle)
$a + 132° = 180°$
so $a = 48°$

b) $b = 45° + 50°$ (exterior angle of triangle)
so $b = 95°$

c) $c + 55° + 90° = 180°$ (angle sum of triangle PQR)
$c + 145° = 180°$
so $c = 35°$

$90° + \text{angle PRS} = 180°$ (angles on a straight line)
$\text{angle PRS} = 90°$
$d + \text{angle PRS} = 145°$ (exterior angle of triangle PRS)
so $d = 55°$

Example 2

Calculate the size of each lettered angle in these sketches.

a) 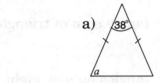   b) 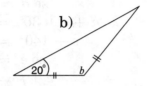   c)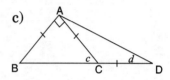

Solution

a) Unmarked angle of the triangle = a (base angles of isosceles triangle)

$a + a + 38° = 180°$ (angle sum of triangle)
$2a + 38° = 180°$
$2a = 142°$
so $a = 71°$

b) Unmarked angle of the triangle = $20°$ (base angles of isosceles triangle)
$b + 20° + 20° = 180°$ (angle sum of triangle)
$b + 40° = 180°$
so $b = 140°$

c) In triangle ABC, AB = AC
so angle B = c (base angles of isosceles triangle)
$c + c + 90° = 180°$ (angle sum of triangle ABC)
$2c = 90°$
so $c = 45°$

In triangle ACD, AC = CD
so angle CAD = d (base angles of isosceles triangle)
$c = d + d$ (exterior angle of triangle ACD)
$45° = 2d$
so $d = 22\tfrac{1}{2}°$

Example 3

Calculate the size of each lettered angle in these sketches.

a) 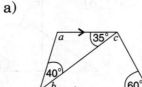   b)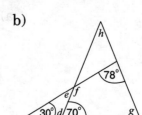

Solution

a) $a + 40° + 35° = 180°$ (angle sum of triangle)
$a + 75° = 180°$
so $a = 105°$
$b = 35°$ (alternate angles)
$b + c + 60° = 180°$ (angle sum of triangle)
$c + 95° = 180°$
so $c = 85°$

b) $d + 70° = 180°$ (angles on a straight line)
so $d = 110°$
$d + e + 30° = 180°$ (angle sum of triangle)
$e + 140° = 180°$
so $e = 40°$
$e + f = 180°$ (angles on a straight line)
so $f = 140°$
$g + 78° + 30° = 180°$ (angle sum of triangle)
$g + 108° = 180°$
so $g = 72°$
$g + h + 70° = 180°$ (angle sum of triangle)
$h + 142° = 180°$
so $h = 38°$

The following problems are for you to solve. Remember that you may have to use any of the facts about angles you previously learnt, as well as the facts about angles of triangles. You must always write which fact you are using, in brackets, when you use it.

EXERCISE 4

1. Calculate the size of each lettered angle in these sketches.

   a) 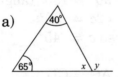   b) 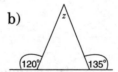   c)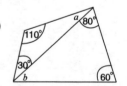

2. Calculate the size of each lettered angle in these sketches.

   a) 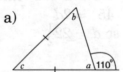   b)    c)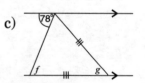

3. Calculate the size of each lettered angle in these sketches.

   a) 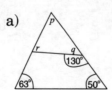   b) 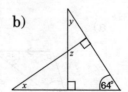   c)

4. In triangle PQR, angle P = 84° and angle Q = 48°.
   a) Calculate the size of angle R.
   b) What special type of triangle is triangle PQR?

Check your answers at the end of this module. Your method could be different from those used there. There are often several correct ways of obtaining the answers to angle calculations – but they all lead to the same final answer! Remember though that if I read your method I must be able to follow what you have said – so you must work one step at a time and *always* give reasons in brackets.

Quadrilaterals

A quadrilateral is a polygon with 4 sides (as its name implies) and so it has 4 vertices and 4 interior angles.

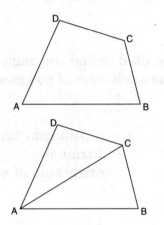

Here is a quadrilateral ABCD.

What is the sum of its 4 interior angles?

You can find the answer to this question by drawing the line from A to C. (This is called a **diagonal** of the quadrilateral. The line from B to D is the second diagonal of ABCD.) Diagonal AC divides the quadrilateral into two triangles, ABC and ADC.

Angle sum of triangle ABC = 180°.
Angle sum of triangle ADC = 180°.

Hence, angle sum of quadrilateral ABCD is 180° + 180° = 360°.

This is another useful result to remember:

> The angle sum of a quadrilateral is 360°.

Example 1

1. Calculate the size of each lettered angle in these sketches.

   a) 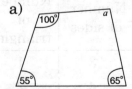   b) 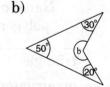   c)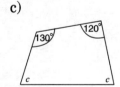

Solution

a) $a + 100° + 55° + 65° = 360°$ (angle sum of quadrilateral)
   $a + 220° = 360°$
   $a = 140°$

b) $b + 30° + 50° + 20° = 360°$ (angle sum of quadrilateral)
   $b + 100° = 360°$
   so $b = 260°$

c) $c + c + 130° + 120° = 360°$ (angle sum of quadrilateral)
$$2c + 250° = 360°$$
$$2c = 110°$$
$$\text{so } c = 55°$$

Example 2

A quadrilateral has angles of $x°$, $2x°$, $(x + 42)°$ and $(2x + 72)°$. Calculate the value of x.

Solution

$x° + 2x° + (x + 42)° + (2x + 72)° = 360°$ (angle sum of quadrilateral)
$$6x + 114° = 360°$$
$$6x = 246°$$
$$x = 41°$$

> The angles of the quadrilateral are 41°, 82°, 83° and 154°. Sum = 360°.

Polygons

The method we used to find the angle sum of a quadrilateral can be used to find the angle sum of polygons with more than 4 sides.

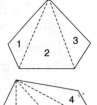

A pentagon can be divided into 3 triangles.
Angle sum of a pentagon = $3 \times 180° = 540°$.

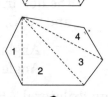

A hexagon can be divided into 4 triangles.
Angle sum of a hexagon = $4 \times 180° = 720°$.

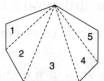

A heptagon can be divided into 5 triangles.
Angle sum of a heptagon = $5 \times 180° = 900°$.

If you put the results of the angle sum calculations in a table, you'll be able to see a pattern and obtain a general formula for the angle sum of a polygon with n sides.

Name of polygon	Number of sides	Number of triangles	Angle sum of polygon
triangle	3	1	$1 \times 180° = 180°$
quadrilateral	4	2	$2 \times 180° = 360°$
pentagon	5	3	$3 \times 180° = 540°$
hexagon	6	4	$4 \times 180° = 720°$
heptagon	7	5	$5 \times 180° = 900°$
octagon	8	6	$6 \times 180° = 1080°$
nonagon	9	7	$7 \times 180° = 1260°$
decagon	10	8	$8 \times 180° = 1440°$
n-agon	n	$n - 2$	$(n - 2) \times 180° = (180n - 360°)$

Don't try to remember the results for particular polygons (except those for the triangle and the quadrilateral). Instead, remember the method we used (dividing the polygon into triangles) and one of these general results:

> The angle sum of a polygon with n sides is $(n - 2) \times 180$ degrees.

or

> The angle sum of a polygon with n sides is $(180n - 360)$ degrees.

Regular polygons

Do you remember that a regular polygon is a polygon with all its angles the same size *and* all its sides the same length?

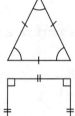

A regular triangle is always called an **equilateral triangle**.

A regular quadrilateral is always called a **square**.

Other regular polygons do not have special names.

You know how to calculate the angle sum of a polygon and, if the polygon is regular, you will be able to find the size of each interior angle, using the fact that all the interior angles are the same size.

Example 1

Find the angle sum of:
a) an octagon
b) a polygon with 20 sides

Solution

a) An octagon has 8 sides, so its angle sum = $8 \times 180° - 360°$
  = $1440° - 360°$
  = $1080°$

b) Angle sum of a 20-sided polygon = $20 \times 180° - 360°$
  = $3600° - 360°$
  = $3240°$

Example 2

Find the size of each interior angle of:
a) a regular decagon
b) a regular 12-sided polygon

Solution

a) A decagon has 10 sides, so its angle sum = 10 × 180° − 360°
    = 1800° − 360°
    = 1440°

Each interior angle of a regular decagon = 1440° ÷ 10
    = 144°

b) A 12-sided polygon has angle sum = 12 × 180° − 360°
    = 2160° − 360°
    = 1800°

Each interior angle of a regular 12-sided polygon
    = 1800° ÷ 12
    = 150°

Example 3

The diagram represents part of a regular octagon ABCD, with the diagonal AC drawn.

a) Calculate angle ABC.
b) Calculate angle ACD.

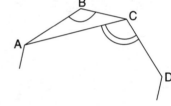

Solution

a) Angle ABC is an interior angle of the regular octagon. An octagon has 8 sides, so angle sum = 8 × 180° − 360° = 1080°.
    The octagon is regular, so each interior angle = 1080° ÷ 8
    = 135°
    Hence, angle ABC = 135°

b) The octagon is regular, so its sides are all equal in length. Hence, AB = BC and so triangle ABC is isosceles.

    In triangle ABC, angle ACB = angle CAB (base angles of isosceles triangle)
    and angle ACB + angle CAB = 180° − 135° (angle sum of triangle)
    = 45°
    It follows that angle ACB = $22\frac{1}{2}°$
    We know that angle BCD = 135° (angle of the regular octagon)
    and so angle ACD = angle BCD − angle ACB
    = 135° − $22\frac{1}{2}°$
    so angle ACD = $112\frac{1}{2}°$

Example 4

Four of the angles of a pentagon are each 100°. Calculate the size of the fifth angle.

Solution

A pentagon has 5 sides, so its angle sum $= 5 \times 180° - 360°$
$= 900° - 360°$
$= 540°$
Sum of the 4 equal angles $= 4 \times 100° = 400°$
Hence, the fifth angle $= 140°$

Example 5

A 7-sided polygon has 2 equal angles of $120°$ and 5 equal angles of $a°$. Find the value of a.

Solution

A 7-sided polygon has angle sum $= 7 \times 180° - 360°$
$= 1260° - 360°$
$= 900°$
Hence, $2 \times 120° + 5a° = 900°$
$240° + 5a = 900°$
$5a = 660°$
so $a = 132°$

Exterior angles of a convex polygon

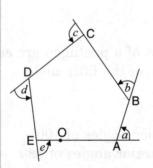

Here is a pentagon ABCDE with its 5 exterior angles (a, b, c, d, e) drawn.

Imagine that you have to walk once round the boundary of the pentagon (anti-clockwise), starting at O and 'turning the corner' as you reach A, B, C, D and E.

When you reach A, you have to turn through angle a so that you can proceed along AB. When you reach B, you have to turn through angle b so that you can proceed along BC. And so on, until you reach O again.

When you have completed your journey, the total angle you have turned through $= a + b + c + d + e$.

However, when you have completed your journey, you are facing in the same direction as when you started, and you have rotated your body through one complete turn.

Hence, $a + b + c + d + e = 360°$.

This argument clearly applies to *any* convex polygon, no matter how many sides it has. We can therefore make this general statement:

> The exterior angles of a convex polygon add up to 360°.

Can you explain why this result is not true for re-entrant polygons? This result gives us another way of calculating angles in polygons.

Example 1

Find the size of each interior angle of:
a) a regular decagon
b) a regular 12-sided polygon

Solution

a) 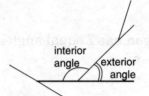 A regular decagon has 10 equal exterior angles which have a sum of 360°.

all regular polygons are convex

Each exterior angle = 360° ÷ 10 = 36°
Each interior angle = 180° − 36°   (angles on a straight line)
Each interior angle = 144°

b) The 12 exterior angles are equal and add up to 360°.
Each exterior angle = 360° ÷ 12 = 30°
Each interior angle = 180° − 30°   (angles on a straight line)
= 150°

Example 2

Four of the angles of a pentagon are each 100°.
Calculate the size of the fifth angle.

Solution

There are 4 interior angles of 100°
so there are 4 exterior angles of 180° − 100°, that is 80°.
These 4 exterior angles add up to 4 × 80°, that is 320°.

But the 5 exterior angles add up to 360°
so the fifth exterior angle = 360° − 320° = 40°.
The fifth interior angle = 180° − 40° = 140°   (angles on a straight line)

Example 3

A 7-sided polygon has 2 equal angles of 120° and 5 equal angles of a°.
Find the value of a.

Solution

The polygon has 2 exterior angles of 180° − 120°, that is 60°.
The 7 exterior angles add up to 360°.
The remaining 5 exterior angles add up to 360° − 2 × 60°,
that is 240°.
Since these 5 exterior angles are equal, each of them is 48°.
Hence, the 5 equal interior angles are each 180° − 48°.
It follows that a = 132°.

Module 4 Unit 1

Example 4

Each of the interior angles of a regular polygon is 165°.
Calculate the number of sides of the polygon.

Solution

Each interior angle = 165°
so each exterior angle = 180° − 165°   (angles on a straight line)
= 15°
The sum of the exterior angles = 360°
so the number of exterior angles = 360° ÷ 15° = 24.
Hence, the number of sides of the polygon = 24.

EXERCISE 5

1. Calculate the size of each lettered angle in these sketches.

   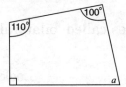 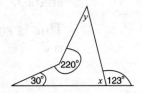

2. A quadrilateral has 3 interior angles each equal to 85°.
   Calculate the size of the fourth interior angle.

3. Find the angle sum of:
   a) an 11-sided polygon
   b) a polygon with 32 sides

4. Find the size of each interior angle of:
   a) a regular pentagon
   b) a regular polygon with 15 sides

5. 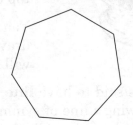 Some countries now have a coin in the shape of a regular 7-sided polygon, as shown in the diagram. Calculate the size of each interior angle of the polygon.
   Give your answer:
   a) as a mixed fraction
   b) to the nearest degree

6. 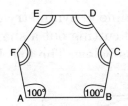 A hexagon ABCDEF has angle FAB = angle ABC = 100°.
   All the other angles are equal.
   Calculate angle BCD.

7. Each interior angle of a regular polygon is 170°. Calculate how many sides the polygon has.

Check your answers at the end of this module.

C Symmetry

John Keats wrote 'A thing of beauty is a joy forever' but another common saying (attributed to Margaret Hungerford) is 'Beauty is in the eye of the beholder'.

When we describe an object as beautiful, some of us would mean that it is balanced and that it shows harmony and regularity of form. Perhaps we would use the word 'symmetrical'.

Two dimensional shapes (those drawn on a flat surface) can have symmetry and so can three dimensional objects (solids such as buildings or human beings). In architecture and art, symmetry may involve colour or texture, but we deal mainly with geometrical symmetry.

Symmetry of solids will be dealt with in Unit 3. In this section we concentrate on the two types of symmetry of two dimensional shapes.

Line symmetry

This is sometimes called 'bilateral symmetry' or 'reflective symmetry'.

Look at this drawing of an African mask. Each half is a reflection, or mirror image, of the other half. If you place a mirror along the dashed line, the half you can see together with its reflection in the mirror is exactly the same as the complete drawing.

Alternatively, if you fold the page along the dashed line, the left-hand half of the drawing will fit exactly on top of the right-hand half.

You could make a tracing of the drawing and turn it over – the tracing will fit exactly on top of the drawing.

The drawing is said to have **line symmetry** and the dashed line is called the drawing's line of symmetry. The **line of symmetry** is sometimes called a 'mirror line'.

Shapes with a line of symmetry can be made by folding a piece of paper once and cutting out a shape based on the fold line, as shown in the diagrams below. The fold line is the line of symmetry of the shape.

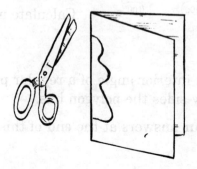 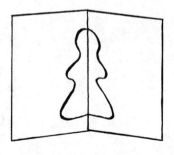

Examples and solutions

1. Each of these shapes has 1 line of symmetry, shown by a dashed line.

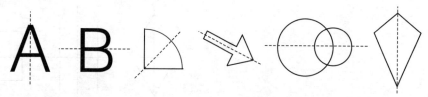

2.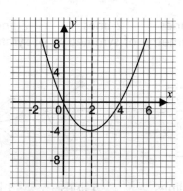

   This is the graph of $y = x^2 - 4x$.

   It has 1 line of symmetry, shown as a dashed line.

   The equation of the line of symmetry is $x = 2$.

3.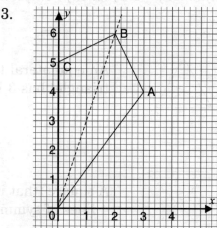

   In this diagram, the points 0 (0,0), A (3,4), B (2,6) and E (0,5) have been plotted.

   The quadrilateral OABC has 1 line of symmetry, shown as a dashed line.

   The equation of the line of symmetry is $y = 3x$.

A two-dimensional shape may have more than one line of symmetry. Shapes with 2 lines of symmetry can be made by folding a piece of paper twice and cutting out a shape across the folds, through 4 layers of paper, as shown in the diagrams below.

Examples and solutions

1. Each of these shapes has more than 1 line of symmetry. In each case, the dashed lines are the lines of symmetry.

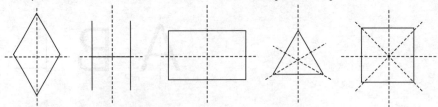

2 lines of symmetry   2 lines of symmetry   2 lines of symmetry   3 lines of symmetry   4 lines of symmetry

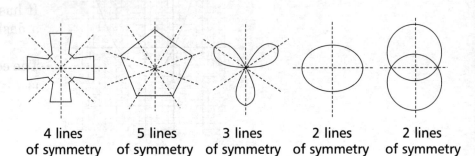

4 lines of symmetry   5 lines of symmetry   3 lines of symmetry   2 lines of symmetry   2 lines of symmetry

2.

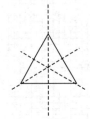

An equilateral triangle (that is a regular triangle) has 3 lines of symmetry.

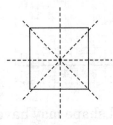

A square (that is a regular quadrilateral) has 4 lines of symmetry.

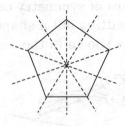

A regular pentagon has 5 lines of symmetry.

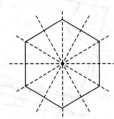

A regular hexagon has 6 lines of symmetry.

In general, a regular n-sided polygon has n lines of symmetry.

Module 4 Unit 1   33

3. Some shapes do not have any lines of symmetry. Here are some examples.

EXERCISE 6

1. Look at the following shapes. For each shape, state how many lines of symmetry it has.

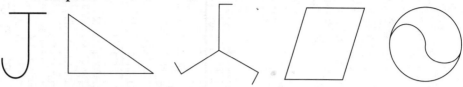

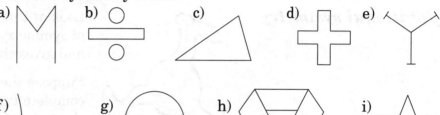

2. Complete these diagrams so that, in each case, the dashed line is a line of symmetry.

3. Complete these diagrams so that, in each case, the dashed lines are lines of symmetry.

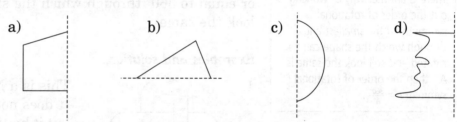

4.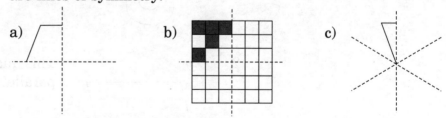

This is the graph of $y = 3x - x^2$.
a) Draw the line of symmetry of the graph.
b) Write down the equation of the line of symmetry.

5.

A quadrilateral has vertices A (1,0), B (4,3), C (3,4) and D (0,1).
a) On the grid, draw the quadrilateral.
b) Draw the 2 lines of symmetry of this quadrilateral.
c) Write down the equations of the lines of symmetry.

Check your answers at the end of this module.

Rotational symmetry

Look at this shape. It does not have any lines of symmetry and yet it has regularity of form and gives the impression of being balanced.

Suppose the shape is rotated through one complete turn about its centre. There are *three* occasions when it looks the same as it did in its starting position. These are when it has been rotated through 120°, 240° and 360°.

We say that the shape has **rotational symmetry** of **order 3**.

The order of rotational symmetry = the number of angles less than or equal to 360° through which the shape can be rotated and still look the same.

> There is another way of working out the order of rotational symmetry. If the *smallest* angle through which the shape can be rotated and still look the same is A°, then the order of rotational symmetry = $\frac{360°}{A°}$.

Examples and solutions

1.

This is a Nigerian wise man's knot. It does not have any lines of symmetry but it has rotational symmetry of order 4. It can be rotated through 90°, 180°, 270° or 360° and still look the same.

the order = $\frac{360°}{90°}$ = 4

2.

This quadrilateral has two pairs of parallel sides. it is a parallelogram

It does not have any lines of symmetry but it has rotational symmetry of order 2. It can be rotated through 180° or 360° and still look the same.

the order = $\frac{360°}{180°}$ = 2

3.

This is a square. It has line symmetry and rotational symmetry.

As you saw previously, a square has 4 lines of symmetry.

The rotational symmetry is of order 4. The square can be rotated through 90°, 180°, 270° or 360° and still look the same.

the order = $\frac{360°}{90°}$ = 4

4.

This is a regular pentagon.

It has 5 lines of symmetry.

It has rotational symmetry of order 5. It can be rotated through 72°, 144°, 216°, 288° or 360° and still look the same.

the order = $\frac{360°}{72°}$ = 5

5.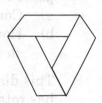

This shape has no lines of symmetry.

It has rotational symmetry of order 3. It can be rotated through 120°, 240° or 360° and still look the same.

the order = $\frac{360°}{120°}$ = 3

6.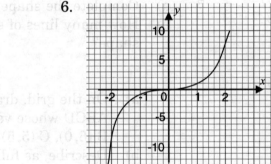

This is the graph of $y = x^3$.

It has no lines of symmetry.

It has rotational symmetry of order 2.

The centre of rotational symmetry is the origin $(0, 0)$.

7.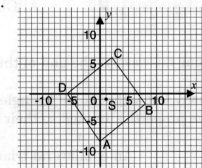

In this diagram, the points A $(0, -8)$, B $(8, -2)$, C $(2, 6)$ and D $(-6, 0)$ have been plotted.

The figure ABCD is a square. It has 4 lines of symmetry.

It has rotational symmetry of order 4 and the centre of rotational symmetry is the point S $(1, -1)$.

You now know how to recognise the two types of symmetry of two dimensional shapes. The following questions will test your understanding of both types.

EXERCISE 7

1. Look at the following shapes and, for each one, answer these questions:
   (i) How many lines of symmetry does the shape have?
   (ii) Does the shape have rotational symmetry? If so, what is the order of rotational symmetry?

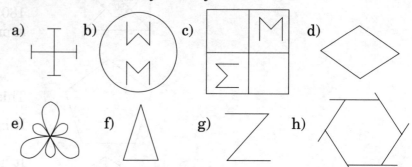

2. This diagram shows part of a shape which has rotational symmetry of order 4 about the point C.
   a) Complete the shape.
   b) How many lines of symmetry has the shape?

3. 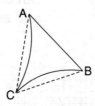 This diagram shows part of a shape which has rotational symmetry of order 5 about the point C. Angle ACB = 72°.
   a) Complete the shape.
   b) How many lines of symmetry has the shape?

4.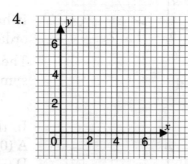
   a) On the grid, draw the quadrilateral ABCD whose vertices are A (1, 1), B (6, 0), C (5, 5) and D (0, 6).
   b) Describe, as fully as possible, the symmetry of the quadrilateral ABCD.

Check your answers at the end of this module.

Special quadrilaterals

You already know that some triangles have special names and that they have properties related to their symmetry.

 An equilateral triangle has 3 lines of symmetry and rotational symmetry of order 3.

 An isosceles triangle has 1 line of symmetry but does not have rotational symmetry.

 A scalene triangle has all three sides of different lengths. It does not have any symmetry.

Bisect means 'cut into two equal parts' so, for example, 'the diagonals bisect each other' means that each diagonal is cut into two equal pieces by the other diagonal.

In a similar way, there are special names given to special types of quadrilateral. These quadrilaterals have properties concerning their sides, their angles and/or their diagonals. Many of these properties are directly related to the symmetry of the quadrilateral concerned.

You will find the names and properties you need to know in the table below.

Name and main property	Lines of symmetry	Rotational symmetry	Other Properties		
trapezium — One pair of sides parallel	None	None			
isosceles trapezium — One pair of sides parallel and the other pair equal	1	None	Two pairs of adjacent angles equal	Diagonals equal (AC = BD)	
kite — Two pairs of adjacent sides equal	1	None	One pair of opposite angles equal	Diagonals at right angles	One diagonal bisects the other and bisects angles
parallelogram — Opposite sides parallel	None	Order 2	Opposite sides equal	Opposite angles equal	Diagonals bisect each other
rhombus — A parallelogram with all its sides equal	2	Order 2	Opposite angles equal	Diagonals at right angles	Diagonals bisect each other and bisect angles

rectangle A parallelogram with all its angles 90°	2	Order 2	Opposite sides equal	Diagonals equal and bisect each other	
square A parallelogram with all its sides equal and all its angles 90°	4	Order 4	Diagonals at right angles	Diagonals equal and bisect each other	Diagonals bisect angles at vertices

You will be expected to use the properties of these special quadrilaterals in answering questions, so learn them well. Here are a few questions for you to try.

EXERCISE 8

1. Each of the following statements can be applied to one or more of the following special quadrilaterals:

   trapezium, isosceles trapezium, kite, parallelogram, rhombus, rectangle, square.

   Write down the name(s) of the quadrilateral(s) to which each statement can be applied.
   a) All the sides are equal in length.
   b) The diagonals are equal in length.
   c) All the angles are 90° and the diagonals bisect each other.
   d) None of the angles is equal.
   e) The diagonals are at right angles.
   f) Does not have two pairs of sides equal in length.
   g) Both diagonals bisect the angles at the vertices.
   h) Another shape which has all the properties of a rhombus.
   i) Has a pair of sides which are parallel but not equal in length.
   j) Has just one line of symmetry.

2. PQRS is a kite with angle P = 90° and angle R = 60°.
   a) Calculate the size of the other two angles of the kite.
   b) Calculate the size of angle PQS.

3. ABCD is a rhombus and angle ABD = 55°.
   Calculate the sizes of the four angles of the rhombus.

4. 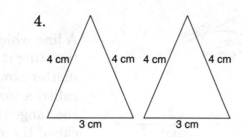 Two pieces of card each have the shape of an isosceles triangle with sides 3 cm, 4 cm and 4 cm.

Show how these two pieces can be fitted together in different ways to form:
a) a rhombus
b) a kite
c) a parallelogram

Check your answers at the end of this module.

D Circles

The Greeks were very interested in circles. Plato regarded a circle as a perfect figure, probably because of its symmetry. Think about the line symmetry and rotational symmetry of a circle. You cannot count the lines of symmetry, they are infinite. Similarly, the order of rotational symmetry of a circle is infinite. This is not true of any other shape.

You can see circles and their uses about you every day. For example, the top of a cup, the wheels on a car and the shape of the sun. You will also have met the circle in your study of mathematics.

 You know that a circle is a shape enclosed by one continuous line called its **circumference**.

All points on the circumference are at an equal distance from a point in the middle of the circle. This point is called the **centre** of the circle.

Here are some other terms and facts which you are expected to know.

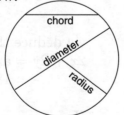 A line from the centre to the circumference is a **radius**. All these **radii** (plural of 'radius') are equal in length.

A line joining two points on the circumference is a **chord**.

A chord which passes through the centre is a **diameter**. The length of a diameter is twice the length of a radius. Every diameter is a line of symmetry of the circle.

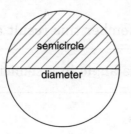 A diameter divides the circle into two equal parts. Each of these parts is called a **semicircle**.

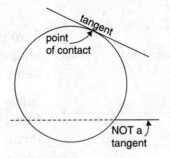

A line which touches the circle (meeting it at one point only, no matter how far it is extended) is called a **tangent**. The point where the tangent touches the circle is called the **point of contact**.

The circle has many useful angle properties. You will study two of these now because they are included in the IGCSE CORE syllabus. The others are in the EXTENDED syllabus and you will look at them in Unit 3.

Angle in a semicircle

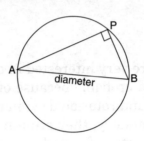

If AB is a diameter of a circle and P is any point on the circumference, then angle APB is a right angle.

This is quite a surprising result, but you can prove it easily by using angle facts which you know already.

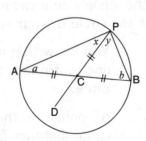

Label the centre of the circle C. Draw the radius PC and extend it to D.

You know that $CA = CP = CB$ (they are radii). Since $CA = CP$, triangle ACP is isosceles and $a = x$.
Since $CP = CB$, triangle PCB is isosceles and $b = y$.

Using
angle $ACD = x + a$ (exterior angle of
         $= 2x$ triangle APC)

angle $DCB = y + b$ (exterior angle of
         $= 2y$ triangle CPB)

we deduce that
$2x + 2y = $ angle ACD + angle DCB
            $= 180°$ (angles on a straight line)

Hence $x + y = 90°$
But angle $APB = x + y$ and so angle $APB = 90°$.

You should remember this result as follows:

> The angle in a semicircle is a right-angle.

Angle between tangent and radius

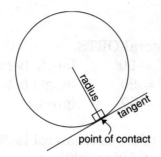

If the radius to the point of contact of a tangent is drawn, then the angle between the radius and the tangent is a right-angle.

This fact follows from the symmetry of the circle. It can be proved as follows:

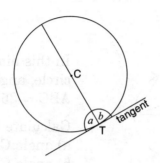

Draw the diameter through the point of contact (T). The diameter passes through the centre of the circle (C).

The diameter is a line of symmetry of the circle.

Hence $a = b$

however, $a + b = 180°$ (angles on a straight line)
and so $a = b = 90°$

Here is the second result about circles that you need to remember:

> The angle between a tangent and radius (through the point of contact) is 90°.

Example 1

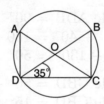

In the diagram, O is the centre of the circle, AC and BD are diameters, and angle BDC = 35°. Calculate the size of angle CAD.

Solution

angle ADC = 90° (angle in a semicircle)
hence angle ADO = 90° − 35° = 55°

OA = OD (radii) and so triangle OAD is isosceles
angle OAD = angle ADO (base angles of isosceles triangle)
= 55°
angle OAD is the same angle as angle CAD and so angle CAD = 55°.

Example 2

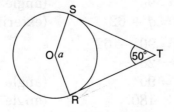

In this diagram, O is the centre of the circle, TR is the tangent to the circle at R, and TS is the tangent to the circle at S. Calculate the size of angle a.

Solution

In the quadrilateral ORTS,
- angle ORT = 90°   (angle between tangent and radius)
- angle OST = 90°   (angle between tangent and radius)
- angle RTS = 50°   (given)

The angle sum of a quadrilateral is 360° and so
$$a + 90° + 90° + 50° = 360°$$
$$a + 230° = 360°$$
$$\text{so } a = 130°$$

Example 3

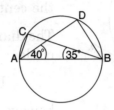

In this diagram, AB is a diameter of the circle, angle DAB = 40° and angle ABC = 35°.

Calculate the size of:
a) angle CAD
b) angle CBD

Solution

a)
$$\text{angle ACB} = 90° \quad \text{(angle in a semicircle)}$$
$$\text{angle CAB} + 90° + 35° = 180° \quad \text{(angle sum of triangle ABC)}$$
$$\text{angle CAB} = 180° - 125° = 55°$$

hence angle CAD = 55° − 40°
$$\text{angle CAD} = 15°$$

b)
$$\text{angle ADB} = 90° \quad \text{(angle in a semicircle)}$$
$$\text{angle DBA} + 90° + 40° = 180° \quad \text{(angle sum of triangle ABD)}$$
$$\text{angle DBA} = 180° - 130° = 50°$$

hence angle CBD = 50° − 35°
$$\text{angle CBD} = 15°$$

Example 4

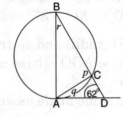

In this diagram, AB is a diameter of the circle and AD is a tangent to the circle.

Calculate the sizes of the angles p, q, r.

Solution

$$p = 90° \quad \text{(angle in a semicircle)}$$
$$p = q + 62° \quad \text{(exterior angle of triangle ACD)}$$
$$\text{hence } q = 90° - 62°$$
$$q = 28°$$
$$\text{angle BAD} = 90° \quad \text{(angle between tangent and radius)}$$
$$90° + 62° + r = 180° \quad \text{(angle sum of triangle ABD)}$$
$$\text{hence } r = 180° - 152°$$
$$r = 28°$$

Example 5

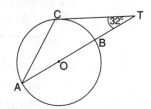

In this diagram, O is the centre of the circle and TC is the tangent to the circle at C.

Calculate the size of angle CAT.

Solution

To solve this problem, I drew the radius OC so that I could use the tangent-radius property.

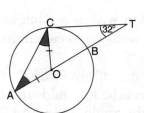

angle OCT = 90°   (angle between tangent and radius)

angle COT = 58°   (angle sum of triangle OCT)

In triangle OAC, OA = OC (radii) so the triangle is isosceles and angle OAC = angle OCA.
But angle COT = angle OAC + angle OCA
 (exterior angle of triangle OAC)
hence angle CAO = 58° ÷ 2 = 29°
angle CAT is the same angle as angle CAO
so angle CAT = 29°

Here are a few questions for you to try. Each of them will require the use of the 'angle in a semicircle' fact or the 'angle between tangent and radius' fact, but you will also have to use some of the other angle facts you have learnt in this module.

EXERCISE 9

1.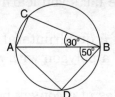

   In this diagram, AB is a diameter of the circle. Calculate the size of angle CAD.

2.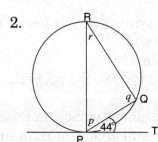

   In this diagram, PR is a diameter of the circle and PT is the tangent at P.

   Calculate the sizes of the angles p, q, r.

3.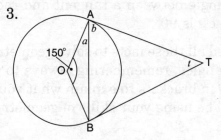

   In this diagram, O is the centre of the circle. TA is the tangent at A and TB is the tangent at B.

   Calculate the sizes of the angles a, b, t.

4.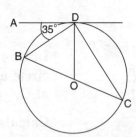

In this diagram, O is the centre of the circle, AD is the tangent at D, and BC is a diameter.

Calculate:
a) angle BDO
b) angle ODC

Check your answers at the end of this module.

Summary

In this unit you started with a section on how to draw, name and measure angles. You went on to calculate the size of angles by using some important facts:
- the angles at a point add up to 360°
- the angles on a straight line add up to 180°
- the angles in a right angle add up to 90°
- vertically opposite angles are equal.

You also learnt some facts to do with parallel lines:
- corresponding angles between parallel lines are equal
- alternate angles between parallel lines are equal
- co-interior angles between parallel lines add up to 180°.

The facts you learnt about triangles are:
- the angle sum of a triangle is 180°
- each exterior angle of a triangle is equal to the sum of the two opposite interior angles
- the base angles of an isosceles triangle are equal
- each angle of an equilateral triangle is 60°.

There are also some facts you learnt about quadrilaterals and other polygons:
- the angle sum of a quadrilateral is 360°
- the angle sum of a polygon with n sides is $(n - 2) \times 180$ degrees or $(180n - 360)$ degrees
- the exterior angles of a convex polygon add up to 360°.

In the section on symmetry you learnt about:
- line symmetry and
- rotational symmetry

and I summarised for you the properties of different quadrilaterals.

Finally you learnt about circles and these facts:
- the angle in a semicircle is a right-angle
- the angle between a tangent and radius (through the point of contact) is 90°.

You used all these facts to solve geometry problems and calculate the size of angles, remembering always to work step by step with reasons in brackets to explain what you were doing. In the next unit you will be using your skills of geometry to solve practical problems.

You should now be ready to attempt the 'Check your progress' which contains questions on the work you have done in Unit 1. These questions are similar to those you will be given in the IGCSE examinations.

Check your progress

1. 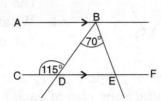 In this diagram, the lines AB and CDEF are parallel. Calculate:
   a) angle BDE
   b) angle ABD
   c) angle BEF

2. 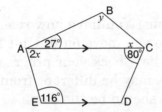 In this diagram, AC is parallel to ED. Angle CAE is twice the size of angle ACB, angle BAC = 27°, angle AED = 116°, angle ACD = 80°.

   Work out the values of x and y.

3. 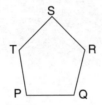 PQRST is a regular pentagon.
   a) Calculate angle PQR.
   b) Calculate angle PQS.

4.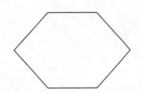
   a) What is the order of rotational symmetry of a regular hexagon, as shown in the diagram?
   b) Add one line to the diagram so that it has rotational symmetry of order 2.
   c) How many lines of symmetry does your diagram have?

5. 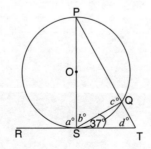 RST is a tangent to the circle, centre O. PS is a diameter. Q is a point on the circumference and PQT is a straight line. Angle QST = 37°.

   Write down the values of a, b, c and d.

6.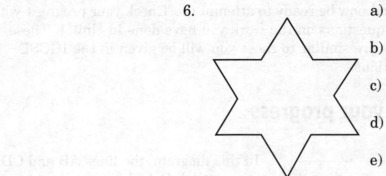
   a) Is this star shape a regular polygon? Explain your answer.
   b) Calculate the angle sum of a 12-sided polygon.
   c) Calculate the size of each of the interior angles of the star shape.
   d) How many lines of symmetry has the star shape?
   e) What is the order of rotational symmetry of the star shape?

7. Each interior angle of a regular polygon is 175°.
   a) Calculate the size of each exterior angle.
   b) Calculate the number of sides of the polygon.

Congratulations! You have now reached the end of this rather long Unit 1. Before you move on to Unit 2, check your answers to the questions in the 'Check your progress' at the end of this module. Your methods may be different from those shown there, but your final answers should be the same.

Unit 2
Practical Geometry

In this unit you will learn how to draw accurate and scale drawings using geometrical instruments. I will also introduce you to the interesting idea of loci and by the end of the unit you should have an understanding of locus and be able to draw the locus of a point.

There are four sections in this unit:

Section	Title	Time
A	Using geometrical instruments	$2\frac{1}{2}$ hours
B	Scale drawings	$2\frac{1}{2}$ hours
C	Geometrical constructions	1 hour
D	Loci	2 hours

By the end of this unit, you should be able to:
- use geometrical instruments to draw shapes accurately
- draw and interpret scale drawings
- understand and use 3 figure bearings
- carry out constructions using only a straight-edge and compass
- understand the term 'locus' and draw loci.

A Using geometrical instruments

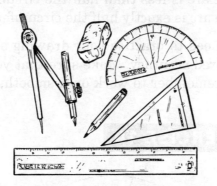

In the IGCSE examinations you may be asked to make an accurate drawing or to take measurements from a drawing which is given in the question.

You must have the right equipment to do this:

- a ruler (to measure lines in cm and mm)
- a protractor (to measure angles)
- a compass with a sharp HB pencil (to draw circles)
- a set square (to draw parallel lines and perpendicular lines)
- another sharp HB pencil
- a rubber (to rub out incorrect work).

For most drawings, you will use a ruler and, for many of them, you will need to use a protractor. Make sure that your ruler and protractor are in good condition – all the markings on them must be clear.

Your drawings must be neat and accurate. You will be expected to draw and measure lengths to the nearest millimetre, and to draw and measure angles to the nearest degree. Remember that small errors tend to build up into big errors.

Always use pencil for drawings, so that if you make a mistake, you can correct it and still have a neat result.

You must never use guesswork in drawing. For example, any line you draw *must* be obtained by joining two points already shown in the drawing or by drawing the line through a known point in a known direction.

In Unit 1, Section A, you learnt how to use a protractor. Now I'll show you how to use other geometrical instruments.

Using a compass

To draw a circle, or part of a circle, you use a **compass**.

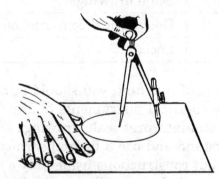

You will need to know the radius of the circle and the position of its centre. The distance between the sharp point of the compass and the pencil point has to be made equal to the radius of the circle. The sharp point of the compass has to be placed where the centre of the circle is to be.

Sometimes only part of a circle needs to be drawn. This is called an **arc**.

If the arc is more than half the circumference, it is a **major arc**.
If the arc is less than half the circumference, it is a **minor arc**.
If the arc is exactly half the circumference, it is a **semicircular arc**.

You need to practise your drawing skills in order to obtain the required accuracy. Make sure that your compass is reasonably tight and remember to work on a smooth, flat surface.

EXERCISE 10

1. 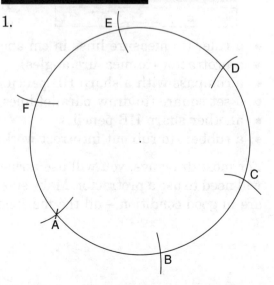 Draw a circle with a radius of 3 cm and mark a point A on its circumference. Using point A as centre, draw an arc with radius 3 cm to cut the circle at B.

    Using point B as centre, draw an arc with radius 3 cm to cut the circle at C.

    Repeat this process to obtain the points D, E, F.

When you use point F as centre and draw an arc with radius 3 cm, you will find that the arc passes through the point A where you started. Label the centre of the original circle O. Join A to B, B to O and O to A.

a) What special type of triangle is triangle OAB?
b) What is the size of angle AOB?
c) Explain why the arc with centre F and radius 3 cm passes through the point A.

2.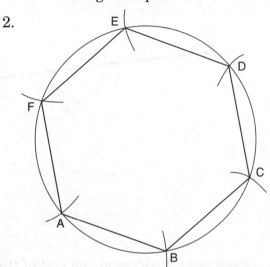

Draw a circle with a radius of 3 cm and obtain the points A, B, C, D, E, F as in question 1. Join A to B, B to C, C to D, D to E, E to F and F to A.

a) What can you say about the lengths of the sides of the figure ABCDEF?
b) Explain why angle ABC is equal to 120° exactly.
c) What is the size of angles BCD, CDE, DEF, EFA and FAB?
d) What special type of figure is ABCDEF?

3.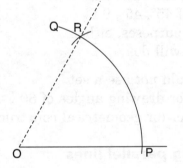

Draw a straight line segment with an end point O. With centre O, draw an arc of a circle to cut the line segment (choose any radius you like). In the diagram, the arc is PQ and it cuts the line segment at P.

Now draw an arc with centre P and the same radius as before. Join O to R, the point where the two arcs cross.

Measure angle POR.
You should find that it is 60°. | this is a method of drawing an angle of 60° without using a protractor

a) Join P to R. What special type of triangle is triangle OPR?
b) Explain why angle POR is exactly 60°.

4.

In this diagram, the outer circle has a radius of 2 cm. The arcs all pass through the centre of the circle and they meet on the circumference of the circle.

a) What is the radius of each of the arcs?

b) Test your skills by drawing the diagram accurately.
c) State whether the arcs are major arcs or minor arcs.
d) How many lines of symmetry has the diagram?
e) What is the order of rotational symmetry of the diagram?

5. 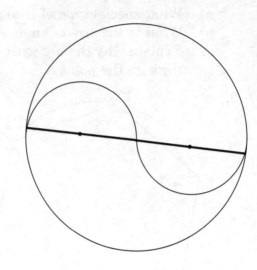 In this diagram, the outer circle has a radius of 3 cm and the two semicircles are the same size as each other.

a) What is the radius of each of the semicircles?
b) Make an accurate copy of the diagram.
c) How many lines of symmetry has this diagram?
d) What is the order of rotational symmetry of the diagram?

Check your answers at the end of this module.

Using a set square

There are two shapes of set square.
One has angles of 30°, 60°, 90°, and the other has angles of 45°, 45°, 90°.
For our purposes, either of these will do.

You should *not* use a set square for drawing angles of 30°, 45° or 60° – you should use a protractor (or geometrical construction) for these angles.

Drawing parallel lines

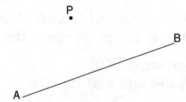

Suppose you have a diagram containing a line AB and a point P, and you want to draw a line through P which is parallel to AB.

First place your set square so that one of its edges XY is along AB.

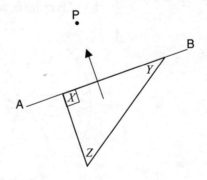

Then place your ruler along another edge XZ. Press the ruler firmly and slide the set square along the ruler until the edge XY passes through P.

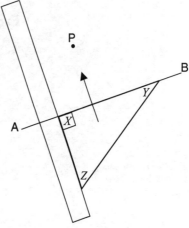

On the paper, draw the line along the edge XY. This is the line through P which is parallel to the line AB.

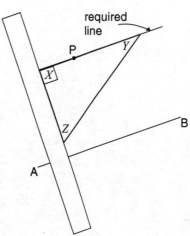

Drawing perpendicular lines

Suppose now that you want to draw a line through the point P which is at right-angles to the line AB.

First place your ruler so that one of its edges is along AB.

Place your set square so that one of its shorter edges XZ is also along AB. Press the ruler firmly and slide the set square along the ruler until the other shorter edge passes through P.

On the paper, draw the line along the edge XZ. This is the line through P which is perpendicular to the line AB.

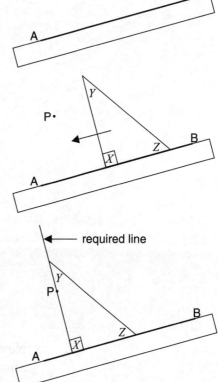

You can also use this method when P is actually on the line AB, that is, to draw the perpendicular to a given line from a point on the line.

Examples and solutions

1. To draw a rectangle which has a length of 4 cm and a breadth of 2 cm, follow these steps:
   - Draw a straight line and on it mark two points, A and B, which are 4 cm apart.
   - Use your set square and ruler to draw the perpendicular to AB at the point A and mark the point D on it so that AD = 2 cm.

   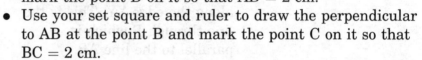

   - Use your set square and ruler to draw the perpendicular to AB at the point B and mark the point C on it so that BC = 2 cm.
   - Join the point C to the point D.

   ABCD is the rectangle required. Check that side DC is parallel to side AB (use your set square and ruler) and that side CD = 4 cm. Remember to show the length and breadth of the rectangle in the diagram.

2. To draw a parallelogram PQRS in which PQ = 4 cm, PS = 3 cm and angle P = 50°, follow these steps:
   - Draw a rough diagram showing the given dimensions.

   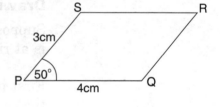

   - Draw a straight line and on it mark two points, P and Q, which are 4 cm apart.
   - Use your protractor to draw a line through P, making an angle of 50° with PQ. On this line, mark the point S which is 3 cm from P.

   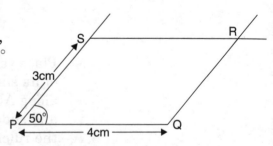

   - Use your set square and ruler to draw:
     (i) the line through S parallel to PQ
     (ii) the line through Q parallel to PS

   Mark the point R where these two lines meet.

   PQRS is the required parallelogram. Check that side SR = 4 cm and that side QR = 3 cm. Make sure the given dimensions are shown in the diagram.

Now test your ability to use a ruler, set square and protractor by answering the following questions.

EXERCISE 11

1. a) Draw a rectangle ABCD in which AB = 4 cm and AD = 3 cm.
   b) Measure the lengths of the diagonals AC and BD.

2. a) Draw a rhombus PQRS in which PQ = 3.5 cm and angle P = 55°.
   b) Measure the lengths of the diagonals PR and QS.
   c) Measure the angle between the diagonals PR and QS.

3. This is a sketch of a trapezium RSTU.
   a) Make an accurate drawing of the trapezium.
   b) Measure the length of the side ST.
   c) Measure the size of angle RST.

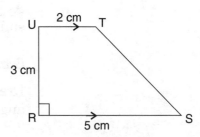

Drawing triangles

Suppose the learners in a class are all asked to draw a triangle with sides of 2 cm, 3 cm and 4 cm. Would you expect their triangles all to be the same shape and size? You may need to rotate the triangles or to turn them over in order to compare them. Can you see that they are the same?

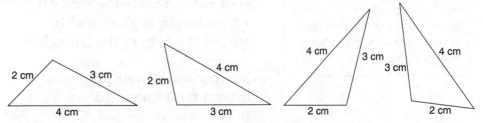

Suppose the learners are now asked to draw a triangle with angles of 50°, 60° and 70°. In this case, the triangles would all be the same shape but not necessarily the same size.

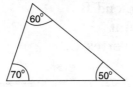 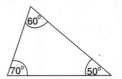

Triangles which are the same shape are said to be **similar**.

Triangles which are the same shape *and* the same size are said to be **congruent**.

Note that, if two triangles are congruent, then they must also be similar. But, if two triangles are similar, they are not necessarily congruent.

> The word similar is not being used in its everyday sense here. In everyday life, 'similar' can mean 'almost the same'. This is very different from 'exactly the same shape but different in size'. To emphasise this, we sometimes use the term 'mathematically similar' instead of 'similar'.

How much information do the learners have to be given before you can be sure that their triangles will all be congruent?

A triangle has 3 sides and 3 angles but you do not need to give the lengths of all the sides and the sizes of all the angles. You will need to give the length of at least one side and if you give the sizes of two of the angles, the size of the third one can be calculated by using the fact that the angle sum of a triangle is always 180°.

Giving the lengths of just two sides is not enough and so is giving the length of one side and the size of one angle.

As you'll see, giving *three* pieces of information is sometimes sufficient, but sometimes you need to give more than this.

Example 1

Draw a triangle ABC in which AB = 6 cm, angle A = 55° and angle B = 45°.

Solution

It is helpful to *sketch* the triangle first. This will help you to decide the order in which you will draw the sides and angles.

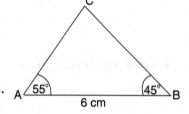

- Start by drawing a 6 cm line with your ruler. This is the side AB whose length is given and it will be the base of the triangle.

> When drawing a triangle, it is usually best to draw the given side (or the longest of the given sides) as the base (that is, the bottom side) of the triangle.

- Making use of your protractor, draw a faint line at 55° to the base at the end A.

- Again using your protractor, draw a faint line at 45° to the base at the end B. Where the two faint lines cross is the vertex C.

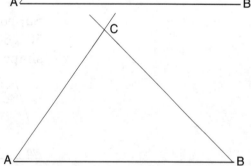

- Draw the lines AC and BC more heavily so that the triangle ABC is clear.

- In the diagram, show the given measurements (6 cm, 55°, 45°).

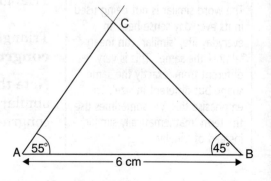

Example 2

Draw a triangle PQR in which QR = 5 cm, angle P = 80° and angle Q = 60°.

Solution

Here is a sketch of the triangle. To draw the triangle accurately, you need to calculate the size of angle R.

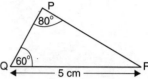

80° + 60° + angle R = 180° (angle sum of a triangle) hence, angle R = 40°.

- Start by drawing a 5 cm line with your ruler. This is the base QR of the triangle.
- Using your protractor, draw a faint line at an angle of 60° to the base at the end Q, and a faint line at an angle of 40° to the base at the end R. Where these two faint lines cross is the vertex P.
- Draw the lines QP and RP more heavily so that the triangle PQR is clear.
- In the diagram, show the given measurements (5 cm, 80°, 60°).

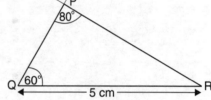

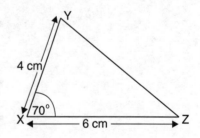

Example 3

Draw a triangle XYZ in which XY = 4 cm, XZ = 6 cm and angle X = 70°.

Solution

Here is a sketch of the triangle. Notice that you are given the size of one angle and the lengths of the arms of the angle. This is sometimes said to be '2 sides and the *included* angle'.

- Start by drawing the longer of the two given sides as the base of the triangle. This is a 6 cm line – the side XZ.
- Using your protractor, draw a faint line at an angle of 70° to the base at the end X.
- Measure accurately 4 cm up this line from X and mark the vertex Y.

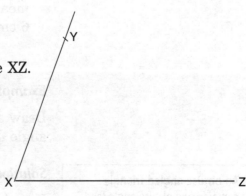

- Using your ruler, join Y to Z and draw the lines more heavily.
- In the diagram, show the given measurements (4 cm, 6 cm, 70°).

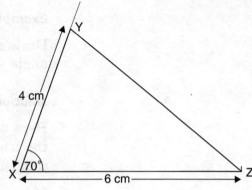

Example 4

Draw a triangle DEF in which DE = 5 cm, EF = 6 cm and FD = 7 cm.

Solution

Here is a sketch of the triangle. It has been drawn so that the longest of the sides is at the bottom.

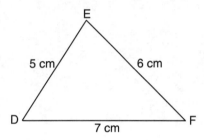

- Start by drawing a 7 cm line with your ruler. This is the base DF of the triangle.
- Now you need your compass. Make the distance between the point and the pencil point equal to 5 cm. With the sharp point positioned at D, draw a faint arc of a circle above the baseline.
- Similarly, draw a faint arc with centre at F and radius 6 cm. Where the two arcs cross is the vertex E.
- Use your ruler to join D to E and F to E.
- In the diagram, show the given measurements (5 cm, 6 cm, 7 cm).

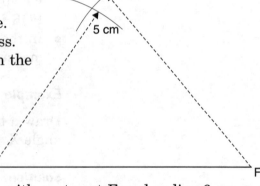

Example 5

Draw a triangle JKL in which JK = 5cm, KL = 6 cm and angle J = 90°.

Solution

Here is a sketch of the triangle. It is called a **right-angled triangle** because it contains a right-angle. (Its other two angles are acute.)

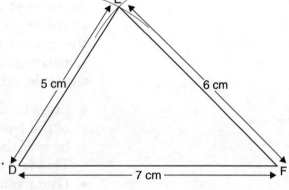

An **obtuse-angled triangle** contains one obtuse angle (and two acute angles).
In an **acute-angled triangle**, all three angles are acute.

- On this occasion it is best to start by drawing the shorter (5 cm) line. This is the side JK of the triangle.
- Using your protractor (or set square), draw a faint line at 90° to JK at the end J.

- Now, using your compass, draw a faint circular arc with centre at K and radius 6 cm. Where the arc crosses the faint line is the vertex L of the triangle.

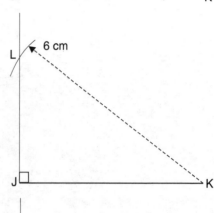

- Using your ruler, join K to L and draw the line JL more heavily.
- In the diagram, show the given measurements (5 cm, 6 cm, 90°).

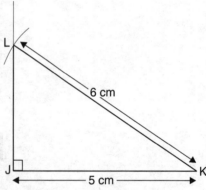

Example 6

Draw a triangle STU in which ST = 6 cm, TU = 4 cm and angle S = 35°.

Solution

Here is a sketch showing the required measurements. Notice that we are given the lengths of two sides and the size of one angle, but the angle is *not* between the two sides. (Compare this with Example 3, where the angle was included.)

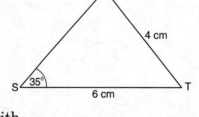

- Start by drawing a 6 cm line with your ruler. This is the base ST of the triangle.
- Using your protractor, draw a faint line at an angle of 35° to the base at the end S.
- Now, using your compass, draw a faint circular arc with centre at T and radius 4 cm. You will find that this arc cuts the faint line at *two* points – these are labelled U_1 and U_2 in the diagram.

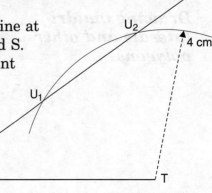

This indicates that there are two triangles (STU_1 and STU_2) which have the given measurements and they are different in shape and size. In other words, they are not congruent.

The two triangles are shown below.

> this is a case where 3 pieces of information about a triangle is *not* enough to fix the shape of the triangle

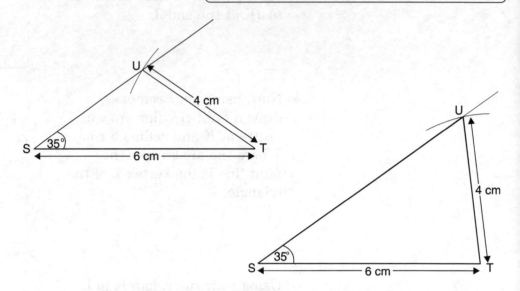

Now you must draw some triangles for yourself. In each of the following questions, you have to draw a triangle and then measure some lengths or some angles. This will be a test of the accuracy of your drawing.

EXERCISE 12

1. Draw a triangle ABC in which AB = 7 cm, angle A = 60° and angle B = 50°. Measure the lengths of the sides BC and CA.

2. Draw a triangle PQR in which QR = 6 cm, angle P = 70° and angle Q = 30°. Measure the lengths of the sides PQ and PR.

3. Draw a triangle XYZ in which XY = 5 cm, XZ = 6 cm and angle X = 60°. Measure angle Y, angle Z and the side YZ.

4. Draw a triangle DEF in which DE = 3 cm, EF = 4 cm and FD = 5 cm. Measure angle D, angle E and angle F.

5. Draw a triangle JKL in which JK = 4 cm, KL = 7 cm and angle J = 90°. Measure angle K and the side JL.

Check your answers at the end of this module.

Drawing quadrilaterals and other polygons

When you have to draw quadrilaterals or other polygons, you will need to use your ruler, protractor, compass and, perhaps, set square, as you did for triangles.

You should proceed as follows:

- Draw a rough sketch showing the lengths of lines and sizes of angles which are given.

- Show on the sketch any measurements which you can deduce using geometrical facts such as: the angles of a quadrilateral add up to 360°, the opposite angles of a parallelogram are equal, co-interior angles between parallel lines add up to 180°, a regular polygon has all its sides equal and all its angles equal, and so on.
- Use the sketch to decide the order in which you are going to draw the lines and angles. There may be more than one possible order – choose the one you think is the easiest.
- Write down the lines and angles in the order you will draw them.
- Take your time in drawing the diagram. Accuracy is of great importance.
- Remember to show the given measurements on your accurate diagram.

Example 1

Draw a quadrilateral ABCD in which AB = 5 cm, AD = 8 cm, CD = 6 cm, angle A = 70° and angle D = 80°.

Solution

Here is a sketch of the quadrilateral. The steps we shall take to draw the quadrilateral are shown in the sketch. There is more than one possible order for these steps. The first step could be 'draw AB' or 'draw CD' instead of the 'draw AD' which I have chosen.

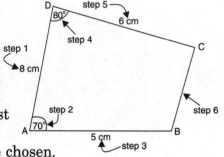

The sides and angles in this accurate diagram were drawn in this order:

Step 1   AD = 8 cm
Step 2   angle A = 70°
Step 3   AB = 5 cm
Step 4   angle D = 80°
Step 5   CD = 6 cm
Step 6   join B to C

Example 2

Draw a regular pentagon with sides of length 4 cm.

Solution

Before you can draw a regular pentagon, you need to know the size of each angle. A pentagon has 5 sides so its angle
sum = $5 \times 180° - 360° = 540°$
(see Unit 1, Section B).

So, each angle of a regular pentagon = $540° \div 5 = 108°$.

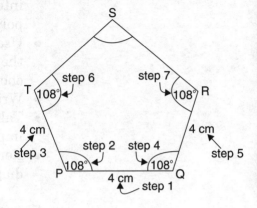

The sketch shows a regular pentagon PQRST with sides of length 4 cm.

The sides and angles in this accurate diagram were drawn in this order:

Step 1 PQ = 4 cm
Step 2 angle P = 108°
Step 3 PT = 4 cm
Step 4 angle Q = 108°
Step 5 QR = 4 cm
Step 6 angle T = 108°
Step 7 angle R = 108°
Step 8 mark vertex S

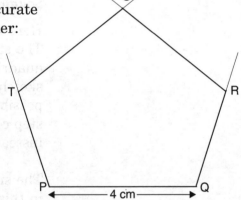

You can check the accuracy of the drawing by measuring TS and SR (they should each be 4 cm) and angle S (it should be 108°).

Example 3

Draw a convex quadrilateral PQRS in which PQ = 6 cm, QR = 4 cm, RS = 5 cm , SP = 3 cm and diagonal PR = 5 cm. Measure the angles of the quadrilateral.

Solution

Here is a sketch of the quadrilateral.

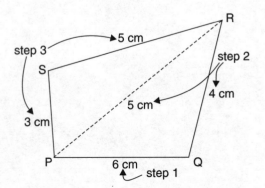

The accurate diagram will be drawn as follows:

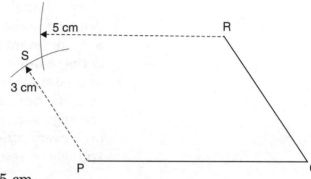

- With a ruler, draw side PQ (6 cm long).
- With a compass, draw a circular arc, centre P and radius 5 cm, and another circular arc, centre Q and radius 4 cm.
- Where the two arcs cross is the vertex R. Join Q to R.
- With a compass, draw a circular arc, centre P and radius 3 cm, and another circular arc, centre R and radius 5 cm.
- Where the two arcs cross is the vertex S. Join S to R and S to P.

the construction arcs must be visible in the accurate diagram

With a protractor, measure the angles of the quadrilateral and obtain angle P = 114°, angle Q = 56°, angle R = 118°, angle S = 72°.

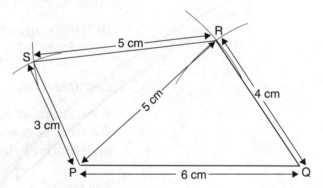

Check
The sum of these four angles is 360°, as it should be – the angle sum of a quadrilateral is 360°.

Here are some questions for you to try.

EXERCISE 13

1. Draw a quadrilateral ABCD in which AB = 6 cm, BC = 4 cm, angle A = 50°, angle B = 120° and angle C = 60°.
   a) Measure the lengths of the sides AD and CD.
   b) What special type of quadrilateral is ABCD?

2. Draw a regular hexagon PQRSTU with sides of length 4 cm. Measure the length of the diagonal PR.

3. Draw a rectangle EFGH with side EF = 5 cm and diagonal EG = 7 cm. Measure the length of the side FG.

Check your answers at the end of this module.

B Scale drawings

So far, we have considered geometrical shapes and drawn them full size. In everyday life, we often have to draw a diagram on a piece of paper to represent something which is much larger or much smaller – for example, a plan of a building or a map of a country or the design of a microchip. Such a diagram is called a **scale drawing**.

The lines in the drawing are all the same fraction of the lines they represent. This fraction is called the **scale** of the drawing.

Every triangle, every rectangle (and every other polygon) in the drawing is the same shape as the one it represents, although they are a different size – the drawing is a miniature (or an enlargement) of the real-life situation. The lines in the scale drawing are all drawn to the same scale, so they are all different in length from the lines in real life, but the angles in the scale drawing are exactly the same as the angles in real life. In other words, every triangle, every rectangle (and every other polygon) is **mathematically similar** to the triangle or rectangle (or polygon) in real life.

Scales

The scale of a diagram, or a map, may be given as a fraction or as a ratio, such as 1/50 000 or 1:50 000.

A scale of 1/50 000 means that every line in the diagram has a length which is 1/50 000 of the length of the line it represents in real life.

Hence, 1 cm in the diagram represents 50 000 cm in real life. In other words, 1 cm represents 500 m or 2 cm represents 1 km.

In IGCSE questions, scales are given in a form such as 1 cm represents 2 km or 1 cm to 2 km. We shall use this form in our work.

Example 1

A rectangular field is 100 m long and 45 m wide. A scale drawing of the field is made with a scale of 1 cm to 10 m. What are the length and width of the field in the drawing?

Solution

10 m is represented by 1 cm, so 100 m is represented by (100 ÷ 10) cm and 45 m is represented by (45 ÷ 10) cm. In the drawing, the length and width of the field are 10 cm and 4.5 cm respectively.

Example 2

On a map, the distance between two villages is 8.4 cm. The scale of the map is 1 cm to 5 km. What is the actual distance between the villages?

Solution

1 cm on the map represents 5 km, so 8.4 cm represents 8.4 × 5 km, that is 42 km. The actual distance between the villages is 42 km.

Example 3

On a map, the distance between Kimberley and Bloemfontein is 7.3 cm. The actual distance between Kimberley and Bloemfontein is 147 km. What is the scale of the map?

Solution

7.3 cm on the map represents 147 km, so 1 cm represents (147 ÷ 7.3) km, that is 20.14 km.

The distances 7.3 cm and 147 km given in the question are only approximate so the result 20.14 km is only approximate. It would not be sensible to say that the scale of the map is 1 cm to 20.14 km. It is much more likely to be 1 cm to 20 km.

The scale of the map is 1 cm to 20 km.

Example 4

A triangular field has sides of 25 m, 39 m, 56 m and angles of 23°, 37°, 120°. A scale drawing is made of the field using a scale of 1 cm to 5 m. What are the lengths of the sides and the sizes of the angles of the field on the drawing?

Solution

5 m is represented by 1 cm on the drawing. Hence, 25 m is represented by (25 ÷ 5) cm, that is 5 cm, 39 m is represented by (39 ÷ 5) cm, that is 7.8 cm, 56 m is represented by (56 ÷ 5) cm, that is 11.2 cm. The angles on the scale drawing are exactly the same as the angles of the actual field. So, on the drawing, the triangular field has sides of 5 cm, 7.8 cm, 11.2 cm and angles of 23°, 37°, 120°.

Test your understanding of scales by answering the following questions.

EXERCISE 14

1. On the plan of a house the living room is 3.4 cm long and 2.6 cm wide. The scale of the plan is 1 cm to 2 m. Calculate the actual length and width of the room.

2. The actual distance between two villages is 12 km. Calculate the distance between the villages on a map whose scale is:
   a) 1 cm to 4 km
   b) 1 cm to 5 km

3. A car ramp is 28 m long and makes an angle of 15° with the horizontal. A scale drawing is to be made of the ramp using a scale of 1 cm to 5 m.
   a) How long will the ramp be on the drawing?
   b) What angle will the ramp make with the horizontal on the drawing?

Check your answers at the end of this module.

Angle of elevation and angle of depression

Scale drawing questions often involve the observation of objects which are higher than you or lower than you, for example, the top of a building, an aeroplane, or a ship in a harbour.

In these cases, the terms **angle of elevation** and **angle of depression** may be used. These are the angles through which you have to turn your eye, starting from the horizontal line of sight towards the object, to look directly at the object. It is important to remember that:

> Angles of elevation and depression are *always* measured from the *horizontal*.

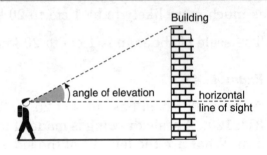

The angle of elevation is measured *upwards* from the horizontal line of sight.

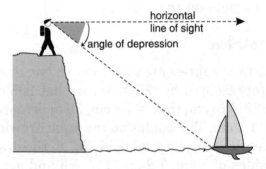

The angle of depression is measured *downwards* from the horizontal line of sight.

Drawing diagrams to scale

You are now ready to solve some problems by scale drawing. You should follow these steps:

- Draw a rough sketch, showing the lengths of lines and sizes of angles which are given.
- When you are told to use a particular scale, use it. Otherwise, choose a suitable scale to use, one that:
  a) is easy to use – for example, one that has a number such as 1, 2, 5, 10, . . . in it (these numbers are easy to divide by),
  b) makes the drawing as large as possible and still fit on the paper (small diagrams will not give very accurate answers).
- Use the scale to work out the lengths of the lines to be drawn in the scale drawing. (Be careful with the units.)
- Make a clean, tidy and accurate scale drawing using appropriate geometrical instruments. Show on it the *given* lengths and angles. Write the scale next to the drawing.
- Measure lengths and angles in the drawing to find the answers to the problem. Remember to change the lengths to *full size* using the scale. Remember that the *full size* angles are the same as the angles in the scale drawing.

Example 1

A vertical radio mast stands on horizontal ground. From a point 50 m from the foot of the mast, the angle of elevation of the top of the mast is 30°. Using a scale of 1 cm to 5 m, make a scale drawing to find the height of the mast.

Solution

Here is a rough sketch. The mast is vertical and the ground is horizontal, so the angle between them is 90°.

On a scale of 1 cm to 5 m, the distance 50 m will be represented by a line of length 10 cm.

The lines and angles in the scale drawing are drawn in this order:

Step 1 AB = 10 cm
Step 2 angle ABC = 90°
Step 3 angle BAC = 30°

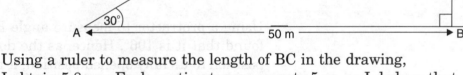

Using a ruler to measure the length of BC in the drawing, I obtain 5.8 cm. Each centimetre represents 5 m, so I deduce that the height of the mast = 5.8 × 5 m = 29 m.

Example 2

This sketch represents a door into an office. The door can be opened until it touches the side wall. Use a scale drawing to find the size of the angle through which the door can be turned from the closed position to the fully opened position.

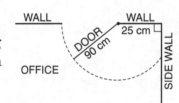

Solution

In the scale drawing, I need to represent a distance of just over (90 + 25) cm from left to right and a distance of just over 90 cm from top to bottom. A sensible scale would be 1 cm to 10 cm (giving a scale drawing of about 11.5 cm by 9 cm).

In the scale drawing, the door will be 9 cm wide and the door hinge will be 2.5 cm from the side wall. The steps in obtaining the drawing are as follows:

- Draw two lines AB and AC at right-angles to each other.
- Mark the hinge H, which is 2.5 cm from A.
- Using a compass, draw a circular arc with centre H and radius 9 cm. (This represents the path of the outer edge of the door as the door is opened.)
- Mark the point D where the arc meets the line AC.

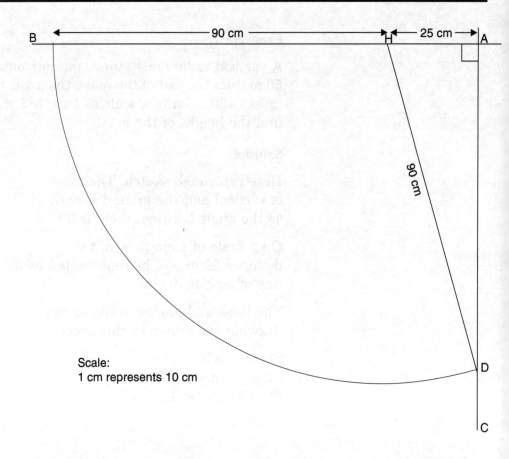

Scale:
1 cm represents 10 cm

Using a protractor to measure angle BHD in the scale drawing, I found that it is 106°. Hence, as the door is moved from its closed position to its fully opened position, it turns through an angle of 106° (the same as in the scale drawing).

Example 3

From the top of a vertical cliff, the angle of depression of a boat is 29°. The top of the cliff is 30 m above sea-level. Using a scale of 1 cm to 5 m, make a scale drawing to find out how far the boat is from the foot of the cliff.

Solution

In this rough sketch,
B represents the boat,
T represents the top of
the cliff and F represents
the foot of the cliff.

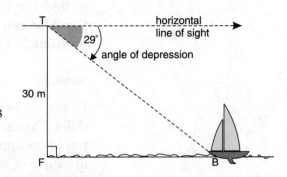

Since the sea is horizontal, angle TBF = 29° (alternate angle to the angle of depression) and angle TFB = 90° (cliff is vertical).
Hence, angle FTB = 61° (angle sum of triangle FTB).

In the scale drawing, 5 m is represented by 1 cm, so 30 m is represented by 6 cm. The steps in obtaining the drawing are as follows:

Step 1 Draw lines FB and FT at right-angles to each other.
Step 2 Mark point T so that FT = 6 cm.
Step 3 Draw angle FTB = 61°.

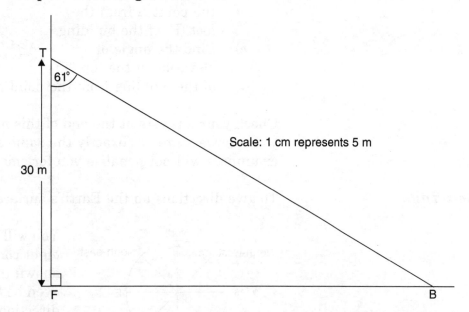

Using a ruler to measure FB in the drawing, I found that it is 10.8 cm. The scale of the drawing is 1 cm to 5 m, so 10.8 cm represents 10.8 × 5 m. Hence, the actual distance between the boat and the foot of the cliff is 54 m.

To obtain sufficient accuracy in scale drawing, you need to be very careful, particularly when you are measuring with your ruler and protractor, and when joining points with straight lines.

See how accurate you can be with the questions in Exercise 15.

EXERCISE 15

1. This is a rough sketch of a field ABCD.
   a) Using a scale of 1 cm to 20 m, make an accurate scale drawing of the field.
   b) Find the sizes of angles C and D at the corners of the field.
   c) Find the length of the side CD of the field.

2. A ladder of length 3.6 m stands on horizontal ground and leans against a vertical wall at an angle of 70° to the horizontal.
   a) What is the size of the angle the ladder makes with the wall?
   b) Use a scale drawing with a scale of 1 cm to 50 cm to find how far the ladder reaches up the wall.

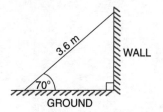

3. This accurate scale drawing represents the vertical wall TF of a building which stands on horizontal ground. It is drawn to a scale of 1 cm to 8 m.

   a) Find the height of the building.
   b) Find the distance of the point A from the foot (F) of the building.
   c) Find the angle of elevation of the top (T) of the building from the point A.

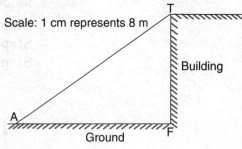

Check your answers at the end of this module. Don't be surprised if your answers are not exactly the same as those given – the IGCSE examiners will not penalise you for small inaccuracies.

Bearings

To give directions on the Earth's surface, you use bearings.

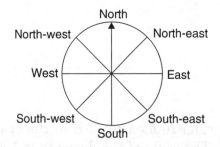

You will already be familiar with the eight compass points which are shown in this diagram. You may even be familiar with other compass directions such as North-north-east, North-north-west, North by East and North by West.

These descriptions are already becoming complicated and, although they may be good enough for giving a general idea of the relative position of points on the Earth's surface, they are not accurate enough, for example, for surveying or for flying an aeroplane.

Directions nowadays are usually given as **3 figure bearings**.
The 3 figure bearing of a point P from an observer O is the angle, measured in degrees in a *clockwise* direction, from the *North* to the line OP. The bearing must contain *3 figures*, so you have to put zeros at the left-hand end if you need to.

For example, 8° must be written as 008°
and 96° must be written as 096°.

This may seem rather complicated but it should be clear after you have looked at the examples below.

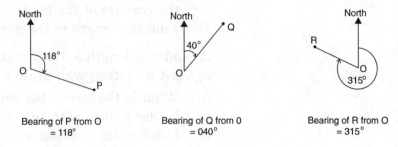

A bearing gives the direction *from* one place to another. Look for the place named immediately after the word 'from' – it is the place at which the angle for the bearing must be measured.

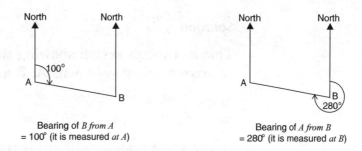

Bearing of B from A = 100° (it is measured *at* A)

Bearing of A from B = 280° (it is measured *at* B)

The North lines through A and B point in the same direction so they are parallel to one another.

Useful fact: For *any* two points P and Q, the difference between the bearing of P from Q and the bearing of Q from P is always 180°. (Can you see why?)

Example 1

Give the 3 figure bearing corresponding to:
a)   West   b)   South-east   c)   North-east

Solution

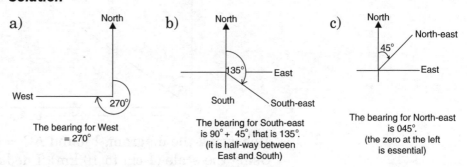

a) The bearing for West = 270°

b) The bearing for South-east is 90° + 45°, that is 135°. (it is half-way between East and South)

c) The bearing for North-east is 045°. (the zero at the left is essential)

Example 2

The bearing of Kimberley from Cape Town is 048°.
What is the bearing of Cape Town from Kimberley?

Solution

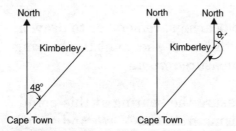

In the second diagram, the two North lines are parallel. Hence, angle $\theta = 48°$ (corresponding angles are equal).

The bearing of Cape Town from Kimberley $= 48° + 180° = 228°$.

Notice that the difference between the two bearings (048° and 228°) is 180°.

Example 3

Town A is 85 km due East of town B. The bearing of town C from town A is 310° and the bearing of town C from town B is 055°.
Using a scale drawing with a scale of 1 cm to 10 km, find:
a)   the distance from town A to town C
b)   the distance from town B to town C

Solution

This is a rough sketch showing the relative positions of towns A, B and C.

angle CBA = 90° − 55° = 35°
angle CAB = 310° − 270° = 40°

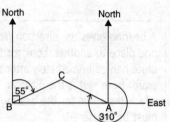

In the scale drawing, 10 km is represented by 1 cm, so 85 km is represented by 8.5 cm and BA will be 8.5 cm long.

The steps in drawing the diagram are as follows:

Step 1 Draw AB 8.5 cm long
Step 2 Draw North lines through A and B (at 90° to AB)
Step 3 Draw BC at 35° to BA (or 55° to North line)
Step 4 Draw AC at 40° to AB (or 50° to North line)

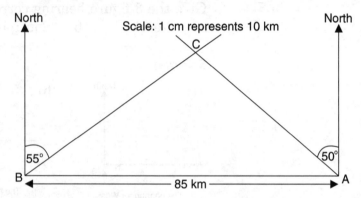

Measuring the diagram, I found AC = 5 cm and BC = 5.6 cm. Using the scale (1 cm to 10 km), I deduced that:

a) the distance from town A to town C = 50 km
b) the distance from town B to town C = 56 km

You should now be ready to tackle some scale drawing problems which involve bearings.

For bearings, remember to draw a North line at the point the bearing is measured *from*.

Measure the bearing at this point, starting from the *North* and going in a *clockwise* direction.

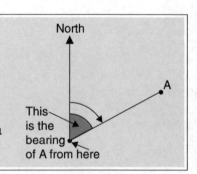

EXERCISE 16

1. Give the 3 figure bearing corresponding to:
   a) East
   b) South-west
   c) North-west

d) North
2. Use the map of southern Africa to find the 3 figure bearing of:
   a) Johannesburg from Windhoek
   b) Johannesburg from Cape Town
   c) Cape Town from Johannesburg
   d) Lusaka from Cape Town
   e) Kimberley from Durban

3. Kimberley is 140 km West and 45 km North of Bloemfontein. Using a scale drawing with a scale of 1 cm to 20 km, find:
   a) the bearing of Bloemfontein from Kimberley
   b) the bearing of Kimberley from Bloemfontein
   c) the direct distance from Bloemfontein to Kimberley

4. Village Q is 7 km from village P on a bearing of 060°. Village R is 5 km from village P on a bearing of 315°. Using a scale drawing with a scale of 1 cm to 1 km, find:
   a) the direct distance from village Q to village R
   b) the bearing of village Q from village R

Check your answers at the end of this module.

C Geometrical constructions

In Section A and Section B, you have learnt how to make accurate drawings using a ruler, protractor, compass and set square. For some drawings, it is not necessary to have all this equipment, and you could regard it as a challenge to draw a diagram with as few simple geometrical instruments as possible.

The Greeks were particularly interested in drawing diagrams using only a straight edge and compass. A straight edge could be the edge of a book but we usually take it to be one edge of a ruler, used only to draw straight lines. We are not allowed to measure with it or to use the top and bottom edges to draw parallel lines.

A drawing made with a straight edge and compass is usually called a **geometrical construction**. When you carry out these constructions, it must be clear that you did not use other instruments, such as a protractor or set square. To do this, you must *not* rub out any circular arcs or straight lines you drew to obtain your diagram.

Drawing an angle of 60°

In Section A you learnt how to draw an angle of 60° without using a protractor. This diagram will remind you how to do it. The diagram is constructed as follows:

- Draw line AB.
- Draw an arc CD (centre A, any radius) cutting AB at C.
- Draw an arc EF (centre C, same radius as arc CD) cutting arc CD at G.
- Draw the line from A to G.
  Then, angle GAB = 60°.

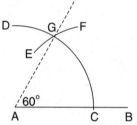

Explanation: Consider the triangle with vertices A, C, G. The sides AC, CG, AG are all the same length (the radius of the two circular arcs). Hence, triangle ACG is equilateral and angle GAB must be 60° (an angle of an equilateral triangle).

Bisecting an angle

'Bisect' means 'cut into two equal parts'. The line which cuts an angle into two equal parts is called the **bisector** of the angle.

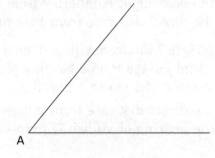

Suppose you want to draw a line which cuts this angle A into two equal parts. You are not allowed to measure the angle with a protractor, halve it and then draw the line! (You may, however, check the accuracy of your construction by using a protractor.)

The method of construction is shown on the following page. Draw an angle and carry out the construction for yourself.

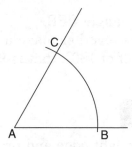 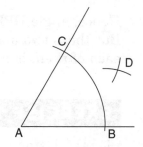 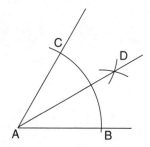

| With centre A, draw a circular arc to cut the arms of the angle at B and C. | Keeping the *same radius* on your compass, draw an arc centre B and an arc centre C. These arcs cross at D. | Draw a straight line from A to pass through D. This line AD bisects the angle BAC. |

Explanation: Consider the quadrilateral with vertices A, C, D, B. It has all four sides (AC, CD, DB, BA) the same length (the radius of the circular arcs). It follows the ACDB is a *rhombus*. The diagonals of a rhombus bisect the angles at the vertices. It follows that angle BAD = angle DAC.

Drawing the perpendicular bisector of a line segment

You already know that 'bisect' means 'cut into two equal parts' and that 'perpendicular' means 'at right angles'.

Suppose you have this line segment AB and that you want to cut it in half with a line which is at right-angles to it.

You are *not* allowed to measure the length of AB or to use a protractor or set square to draw a right-angle.

The method of construction is shown below. Draw a line segment and carry out the construction for yourself.

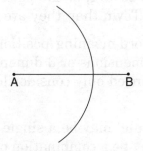 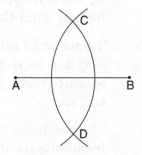 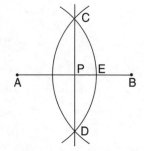

| With centre A, draw a circular arc with radius greater than a half of AB. | Keeping the *same radius*, draw an arc with centre at B. The two arcs cross at C and D. | Draw the straight line through C and D. This line is perpendicular to AB and cuts it into two equal pieces. |

Explanation: The lengths of the sides of the triangle CPE are exactly the same as the lengths of the sides of triangle DPE. It follows that these triangles are congruent, so they have the same sized angles.

Hence, angle DPE = angle CPE.
But these two angles together make a straight angle (180°) and so each of them is a half of 180°, that is 90°.

EXERCISE 17

1. a) Using a straight edge and protractor, draw an angle of 80°.
   b) Using a straight edge and compass only, bisect your angle of 80°.
   c) Use your protractor to check the accuracy of your work.

2. Using a ruler and compass, draw a triangle PQR with PQ = 10 cm, QR = 9 cm and RP = 8 cm. Using a straight edge and compass only, construct the bisectors of the angles P, Q, R of the triangle. What do you notice about these three angle bisectors?

3. Draw a quadrilateral ABCD with AB = 7 cm, AD = 6 cm, angle A = 60°, angle B = 70°, angle C = 120° and angle D = 110°. Using a straight edge and compass only, construct the perpendicular bisectors of the sides AB, BC, CD and DA. What do you notice about these four perpendicular bisectors?

4. Using a straight edge and compass only, draw an angle of 30°.

Check your answers at the end of this module.

D Loci

A **locus** is a set of points which satisfy a given condition (or rule). For example, the set of points in South Africa which are exactly 30 km from Johannesburg, or the places in South Africa which are further from Cape Town than they are from Kimberley.

'Locus' is a Latin word meaning 'position' or 'place'. Its plural is **loci**. Loci can be in 2 dimensions or 3 dimensions but, for the IGCSE examinations, you need only consider loci in a plane (that is, a flat surface).

The condition (or rule) may be a single equality or a single inequality, or it may be a combination of equalities and inequalities. Generally speaking, when the rule is an equality, the locus is a line (it may be straight or curved) and you can imagine that it is the path of a point which is moving according to the rule. In simple cases, you can determine the locus by plotting a number of points which satisfy the rule and joining them up with a straight line or a smooth curve.

When the rule is an inequality, the locus is a region of the plane. Its boundary can be determined by replacing the inequality sign in the rule by an equals sign.

Example 1

The position of a point A is fixed and a point P is to be marked so that AP = 2 cm.

Describe the locus of possible positions of P.

Solution

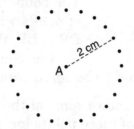

Some of the possible positions of P are marked in this diagram.

The more positions that are marked, the clearer it becomes that the locus of P is the circumference of the circle with centre at A and with radius 2 cm.

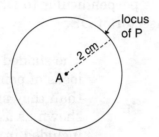

The locus of P is shown in this diagram.

Example 2

The positions of two points, B and C, are fixed. A point P is to be marked so that angle BPC = 90°.

Describe the locus of possible positions of P.

Solution

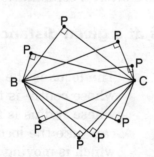

Some of the possible positions of P are shown in this diagram. (Positions of P can be obtained practically using two pins and a set square. The pins are placed at B and C. The set square is placed between the pins so that the arms of the right-angle touch the pins. Mark the positions of the vertex of the right-angle as the set square rotates.)

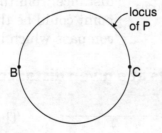

The locus of P is the circumference of the circle which has BC as a diameter.

Example 3

The positions of two points, D and E, are fixed. A point P is to be marked so that it is nearer to D than it is to E.

Describe the locus of possible positions of P.

Solution

```
        .
        .
        .
        .
  D     .     E
        .
        .
        .
        .
```

The rule given is equivalent to the inequality DP < EP.

Because the rule is an *inequality*, the locus is a *region* of the plane.

The boundary of the region is the set of points which satisfy the equality DP = EP, that is the set of points which are the same distance from D as they are from E (they are said to be *equidistant* from D and E).

The diagram above shows some of the points on the boundary. It is clear that the locus of these points includes the mid-point of DE and, by symmetry, it is perpendicular to DE. In other words, it is the perpendicular bisector of DE.

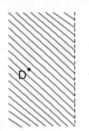

The shaded region in this diagram is the locus of points which are nearer to D than they are to E. (The boundary is shown as a dashed line because it is not included in the region.)

Standard loci

Loci can be obtained by experiment, but this can be time-consuming. It is worthwhile to remember a few simple results which you can obtain by experiment or by common sense. You will find that complicated loci are often a combination of simple loci.

Locus of points at a given distance (d) from a given fixed point (C)

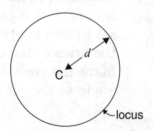

The locus is the circumference of a circle whose centre is the given fixed point (C) and whose radius is the given distance (d). (You can take the locus to be the path of a point which is moving so that it is always at a fixed distance from the fixed point C. The moving point could be the point of the pencil in your compass which has its sharp point at C.)

Locus of points at a given distance (d) from a given straight line (l)

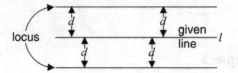

The locus is two straight lines which are both parallel to the given line (l) and a distance d from it.

One part of the locus is 'above' the given line and the other part is 'below' the given line. The given line (l) is of infinite length, but you can only draw part of it. The two parts of the locus are also of infinite length.

Locus of points at a given distance (d) from a given line segment AB

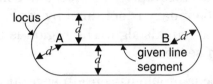

The locus is shown in this diagram. It is in four parts:
a line segment distance d above AB,
a line segment distance d below AB,
a semicircle radius d, centre A,
a semicircle radius d, centre B.

You should convince yourself that every point in the locus is a distance d from the *nearest* point of the line segment. This locus is different from the one above because a line segment has a finite length.

Locus of points equidistant from two given fixed points A and B

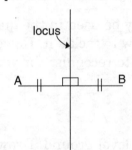

'Equidistant from A and B' means that every point P in the locus is the same distance from A as it is from B, that is PA = PB.

The locus is the perpendicular bisector of the line segment AB.

Locus of points equidistant from the arms of a given angle

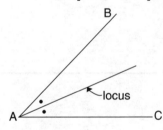

In this diagram, AB and AC are the arms of the given angle. Every point P in the locus is the same perpendicular distance from AB as it is from AC.

The locus is the bisector of the given angle.

Locus of points equidistant from two given intersecting straight lines

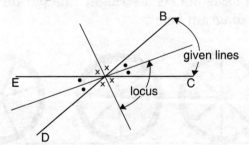

This is similar to the situation above – the arms of the given angle BAC have been extended beyond A.

In the diagram, BD and CE are the given lines. Every point P in the locus is the same perpendicular distance from BD as it is from CE.

The locus consists of the two straight lines which bisect the angles between the given lines.

The two lines which make up the locus are at right-angles to each other. (Can you see why?)

Solving locus problems

For each of the standard loci mentioned above, the rule is an equality – two distances have to be *equal*. In locus problems, the rule may be an inequality. The locus in this case is a *region* of the plane, and its boundary is obtained by replacing the inequality sign in the rule by an equals sign. The region is then obtained by using your common sense.

Some locus problems contain several rules. If this is the case, you must deal with each rule separately and then find the points which obey *all* the rules – that is, find the intersection of the various loci.

You will need to use your geometrical instruments and you may have to draw diagrams to scale. On occasions, you will be told to *construct* a locus. This means that you must use a straight edge (or ruler) and compass only, and you must leave all the construction lines in your diagram.

The word 'locus' may be used in the question, in which case you will have a good idea how to tackle it. However, the word 'locus' may not be used – you have to recognise for yourself that the question is about a locus.

Example 1

A wheel rolls along level ground. Draw the locus of its centre.

Solution

The height of the centre (C) of the wheel above the ground is always equal to the radius of the wheel.

The locus of C is a straight line parallel to the ground, as shown in the diagram below.

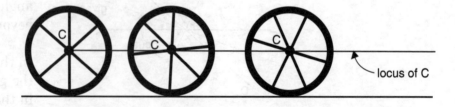

Example 2

a) Using a ruler and compass only, find the point P, inside the rectangle ABCD, which is:
   (i) 4 cm from C and
   (ii) equidistant from AD and AB

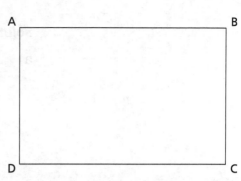

b) Measure and write down the length of BP, to the nearest millimetre.

the word 'locus' is not used in this question but the point P is clearly the intersection of two loci – the locus of points 4 cm from C and the locus of points equidistant from AD and AB.

Solution

a) Locus (i) is part of a circle with centre C and radius 4 cm. (It is not a complete circle because P has to be inside the rectangle ABCD.) Locus (ii) is the bisector of angle BAD. (P has to be equidistant from the arms AD and AB of the angle.)

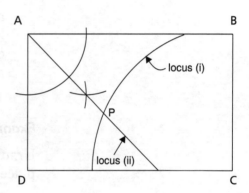

There is only one possible position for the point P – it must be where the angle bisector intersects the quarter circle.

b) Using a ruler to measure the length of BP, I found that it is 42 mm, to the nearest millimetre.

Example 3

The diagram represents a small square field, of side 8 m. It is drawn using a scale of 1 cm to represent 2 m. Pumi ties his donkey to a post, using a rope of length 5 m.

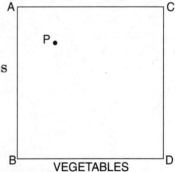

a) One day he fixes the post at the point P shown in the diagram.
   (i) If the rope is fully extended, draw the path that the donkey can walk.
   (ii) Shade, on the diagram, the area of the field in which the donkey can eat.
b) Pumi has planted vegetables outside the field, up to and along the edge BD. He does not want the donkey to get within one metre of them. Show, in a diagram, the area in which Pumi can safely fix the post.

Solution

a) (i) The path that the donkey can walk is the arc of a circle with centre P and radius 5 m. It is shown as locus (i) in this diagram.

(ii) The area in which the donkey can eat is the part of the circle which is shaded in the diagram.

b) The donkey must not get within 1 m of BD and the rope is 5 m long. It follows that the post must be at least 6 m away from BD. The locus of points exactly 6 m from BD is a line parallel to BD and 6 m from it. It is shown as locus (b) in the diagram. The post can be safely fixed anywhere in the area which is shaded in the diagram.

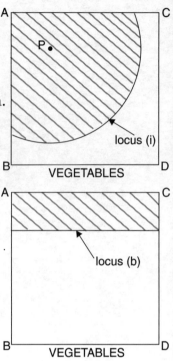

Example 4

Ursula buys a plant to keep in her room. She is told that it must be placed within 2 m of the window.

The diagram below is a plan of her room. Draw accurately and shade the area in which Ursula should place her plant.

Scale:
1 cm to 1 m

Solution

Consider first the points which are exactly 2 m from the window.

> replacing the inequality 'less than 2 m' by the equality 'equal to 2 m'

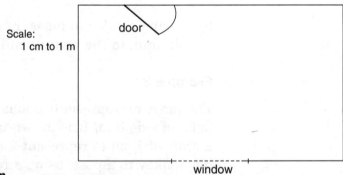

The locus must be inside the room.

The locus consists of the two quarter circles labelled (a) and (b) in the diagram above, and the line segment labelled (c).

The plant must be placed within 2 m of the window. It follows that it must be placed between the window and the locus formed by (a), (b) and (c). This area is shaded in the diagram below.

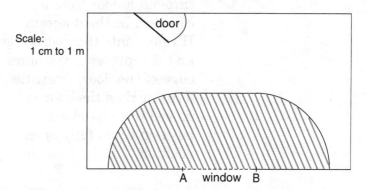

Example 5

a) Draw accurately the locus of points 3 cm from the point A. Label the locus L_1.
b) Draw accurately the locus of points 3 cm from the line ℓ. Label the locus L_2.
c) Indicate, by shading, the locus of points which are at most 3 cm from A and at least 3 cm from ℓ.

Solution

a) Locus L_1 is the circumference of a circle with centre A and radius 3 cm.
b) Locus L_2 is a pair of lines parallel to ℓ. One is 3 cm above ℓ and the other is 3 cm below ℓ.
c) The shaded region contains all the points which are at most 3 cm from A and at least 3 cm from ℓ.

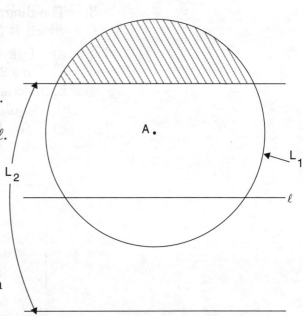

It is time for you to try some locus questions for yourself. Remember that any construction lines you use must be visible in your diagrams.

EXERCISE 18

1. An office can be entered through a door from a corridor. In the diagram, H represents the door hinge and E represents the outer edge of the door. Draw the locus of E as the door is moved from its closed position to its fully open position.

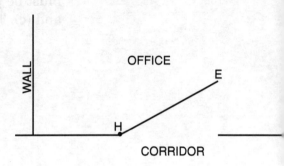

2.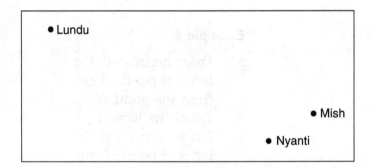

The map above, drawn to a scale of 4 cm to represent 1 km, shows the positions of three villages, Lundu, Mish and Nyanti. Simon's house is the same distance from Mish as it is from Lundu. The house is also less than $\frac{3}{4}$ km from Nyanti. Mark on the map the possible positions of Simon's house.

3. The diagram below is a scale drawing of a rectangular room which is 5 m long and 3 m wide.

   a) Draw accurately the locus of points inside the room which are 3 m from the wall AB.
   b) Construct the locus of points inside the room which are equidistant from the walls AD and DC.
   c) (i) Mark as X the point where the loci intersect.
      (ii) What is the distance, in metres, of X from the wall BC?

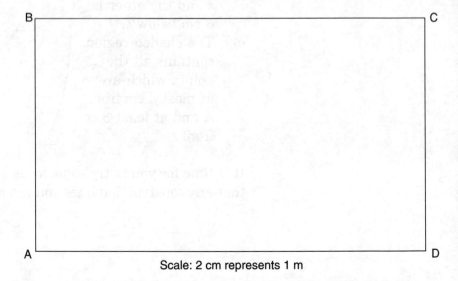

Scale: 2 cm represents 1 m

4.

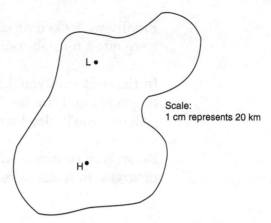

Scale:
1 cm represents 20 km

This is a map of an island, drawn to a scale of 1 cm to represent 20 km. The government has built television transmitters at H and L. The transmitter at H is high powered and has a range of 80 km. The transmitter at L is low powered and has a range of 40 km.

a) On the diagram, show accurately the part of the island where television signals can be received.
b) The government intends to build a third transmitter so that television signals can be received in all parts of the island.
  (i) Will it be necessary to build a high powered transmitter (range 80 km) or will a low powered transmitter (range 40 km) be sufficient?
  (ii) On the diagram, mark with an X a possible position for the third transmitter.

Check your answers at the end of this module.

Summary

Well done! You have now completed the work in Unit 2 and you are well on the way to completing Module 4.
In this unit you learnt how to use geometrical instruments to:
- draw circles – using a compass
- draw parallel and perpendicular lines – using a set square and a ruler
- draw triangles – you learnt about similar triangles and congruent triangles
- draw polygons.

I showed you how to draw a sketch first to work out in which order to work, depending on what information you are given. You also worked with scale drawings and how to represent angles given as:
- angles of elevation or angles of depression
- 3 figure bearings.

In the section on geometrical constructions I showed you how to do the following constructions using *only* a compass and ruler (to draw lines, not measure):
- an angle of 60°
- bisecting an angle
- the perpendicular bisector of a line segment.

Finally we looked at some standard loci and saw that loci problems were often a combination of these standard loci.

In the next unit you'll be learning about solid figures and about congruent and similar figures. If you're doing the IGCSE EXTENDED syllabus you'll also learn more about the geometry of the circle.

Before you continue with the next unit do the following 'Check your progress' to make sure you have mastered the topics in this unit.

Check your progress

1. Using a ruler and compass only, construct:
   a) triangle PQR with PQ = 10 cm, QR = 9 cm and RP = 7 cm
   b) the perpendicular bisectors of PQ and QR
   c) the circle, with centre at the point where the perpendicular bisectors meet, passing through P, Q and R. (This is the **circumcircle** of triangle PQR.)

2. The diagram below shows part of a coastline and two coastguard stations, A and B. A ship in distress sends out a signal. The signal is picked up by both coastguard stations. The bearing of the ship from A is 130°. The bearing of the ship from B is 205°. Use your protractor and ruler to find the position of the ship in the diagram. Mark its position X.

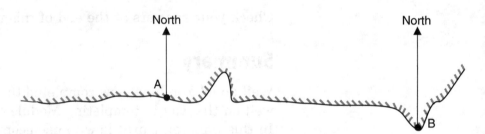

3. The diagram below represents a garden ABCDE.
   AB = 2.5 m, AE = 7 m, ED = 5.2 m, DC = 6.9 m,
   angle EAB = 120°, angle DEA = 90° and angle EDC = 110°.

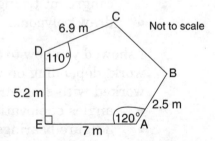

   a) (i) Using a scale of 1 cm to represent 1 m, construct an accurate plan of the garden.
   (ii) Construct the locus of points equidistant from CD and CB.
   (iii) Construct the locus of points 6 m from A.

b) A fountain is to be placed nearer to CD than to CB and no more than 6 m from A.
Shade, and label R, the region within which the fountain could be placed in the garden.

4.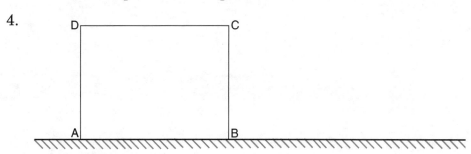

In the scale drawing above, the rectangle ABCD represents a sheet of glass standing vertically on horizontal ground on its edge AB. The sheet of glass is to be rotated in a vertical plane, without slipping, about the point B, until it is standing on the edge BC.

Draw accurately on the diagram:

a) the new position of ABCD, labelling clearly the vertices A, C and D
b) the locus of the point D for this rotation

Check your answers at the end of this module.

Unit 3
More Terms and Facts

This is the final unit of Module 4. In this unit you'll learn about the shapes of objects and their symmetry. You'll also take a closer look at the topic of similar and congruent figures. If you're studying the IGCSE EXTENDED syllabus you'll study congruent triangles in more detail and you'll also learn more about circle geometry.

This unit is divided into three sections:

Section	Title	Time
A	Solid figures	2 hours
B	Congruent figures and similar figures	2 hours
C	More results about circles	3 hours

By the end of this unit, you should be able to:
- name, describe and draw representations of solid figures
- recognise the symmetry of solid figures
- recognise congruent figures and similar figures
- determine whether triangles are similar
- solve problems involving similar triangles
- determine whether triangles are congruent
- solve problems involving congruent triangles
- use the symmetry properties of circles
- use angle properties of circles.

A Solid figures

In everyday life, the solids you handle and the buildings you see come in all shapes and sizes.

To study solids, you can group together those which have some feature in common.

Consider, for example, a tennis ball, a football, a table tennis ball, a golf ball and the moon. Although they differ in size, and their surfaces vary in texture, they have the same general shape.

sphere

The mathematician smooths out the surface and obtains a shape called a **sphere**. This shape has one curved surface and a centre – every point on the surface is the same distance from the centre.

The sphere can be regarded as the 3-dimensional equivalent of a circle.

Polyhedra

You may meet names of particular polyhedra such as 'tetrahedron' (which means 'four faces'), 'octahedron' (which means 'eight faces'), and 'dodecahedron' (which means 'twelve faces'), but these names are not used in IGCSE questions without explanation.

The same word is used for the corner of a solid as is used for the corner of a polygon. The plural of 'vertex' is 'vertices'.

In Unit 1 Section B, you met the term 'polygon' – this is a closed shape, drawn on a flat surface, whose boundary consists of straight lines. The 3-dimensional equivalent of a polygon is a **polyhedron** – this is a solid whose surface consists of polygons. The word 'polyhedron' comes from Greek, meaning 'many faces'. The plural of 'polyhedron' is 'polyhedra'.

The most common polyhedron is a **cube**.

Its surface consists of 6 squares.

Each of these squares is called a **face** of the solid.

Where two faces meet is a straight line called an **edge** of the solid.

A point where edges meet is called a **vertex** of the solid.

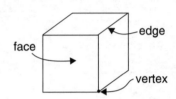

Can you see that a cube has 6 faces, 12 edges and 8 vertices? If not, you should make use of a sugar cube, a cubical dice or a cubical box to help you see this.

A polyhedron whose surface consists of 6 rectangles is called a **cuboid** or a **rectangular block**.

A cuboid has the same number of faces, edges and vertices as a cube. That is, 6 faces, 12 edges and 8 vertices.

A **prism** is a polyhedron which has 2 faces (the top and bottom faces) which are identical polygons, with their corresponding sides parallel. The other faces are usually rectangles, in which case we call the solid a **right prism**.

However, the faces could be parallelograms, in which case the prism is 'leaning over'.

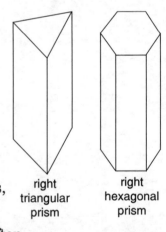

The shape of the top and bottom faces is often used to describe the prism – for example, 'triangular prism', 'hexagonal prism' and 'pentagonal prism'.

A triangular prism has 5 faces, 9 edges and 6 vertices. A hexagonal prism has 8 faces, 18 edges and 12 vertices.

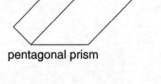

If a prism is cut with a plane parallel to the top and bottom faces, the new faces created are identical to the top and bottom faces. We say that a prism has a 'constant cross-section'.

A **pyramid** is a polyhedron in which the vertices of one face (the base) are joined by edges to one other vertex (the apex). Except perhaps for the base, the faces are all triangles.

triangular pyramid

square based pyramid

When the edges joining the base to the apex are all equal in length, the solid is called a **right pyramid**.

The shape of the base is often used in the description of the solid – for example, 'triangular pyramid'. (A triangular pyramid is also called a 'tetrahedron'.)

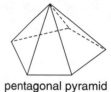
pentagonal pyramid

A triangular pyramid has 4 faces, 6 edges and 4 vertices. A square based pyramid has 5 faces, 8 edges and 5 vertices.

Drawing polyhedra

There are a number of ways of representing a solid by a drawing on a piece of paper. Here are three representations of a cuboid.

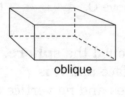
oblique

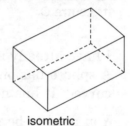
isometric

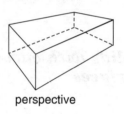
perspective

When you represent a 3-dimensional object by a single 2-dimensional picture, there is bound to be some distortion. In particular, right angles may not be represented by right angles, equal lines may not be represented by equal lines and parallel lines may not be represented by parallel lines. It is not always easy to draw, or to interpret, a representation of a solid, and you will need to develop these skills by practice. You'll usually use oblique drawings. In this type of drawing, parallel lines are represented by parallel lines, and equal lines in any one direction are represented by equal lines (so midpoints are represented by midpoints). Unless the solid is made of transparent material such as glass or perspex, it is usually impossible to see all its edges when viewing it from one direction. In your drawings, you can show the edges which are hidden from view by dashed or dotted lines.

To draw a cuboid

Draw a rectangle. (Make the lines faint.)

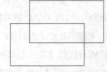
Draw an identical rectangle with sides parallel to sides of the first rectangle.

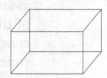
Join the corresponding vertices.

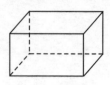
Draw visible edges heavily and hidden edges dashed.

To draw a triangular prism

 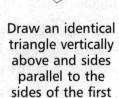

| Draw a triangle. | Draw an identical triangle vertically above and sides parallel to the sides of the first triangle. | Join the corresponding vertices. | Draw visible edges heavily and hidden edges dashed. |

To draw a right, square based pyramid

| Draw a rhombus and mark its centre O. | Mark a point V vertically above O. | Join V to the vertices of the rhombus. | Draw the visible edges heavily and hidden edges dashed. |

Solids with curved surfaces

I have already mentioned the **sphere**. A sphere has only 1 face and it is curved. It has no edges and no vertices.

You need to be familiar with some other solid shapes which have a curved face.

A **cylinder** has a circular base, a circular top of the same size, and a curved face formed by joining the points on the circumferences of the base and top by lines perpendicular to the base.

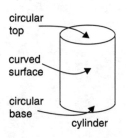

A cylinder has 3 faces (2 plane, 1 curved), 2 edges (curved) and no vertices.

A cylinder can be regarded as being a special sort of prism – a prism with a base which is a regular polygon with a very large number of sides.

A **cone** has a circular base and a curved face formed by joining a single vertex to the points on the circumference of the base. (In our work, we shall only consider 'right circular cones', that is cones in which the line joining the vertex to the centre of the base is perpendicular to the base.)

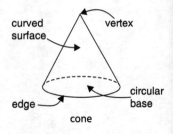

A cone has 2 faces (1 plane, 1 curved), 1 edge (curved) and 1 vertex.

A cone can be regarded as a special sort of pyramid.

Nets

If you have difficulty in interpreting a drawing of a solid, you will find it useful to make a model of the solid.

You may be able to make the surface of the solid by folding a single piece of card. A shape which can be cut out and folded to make the surface of a solid is called a **net** of the solid.

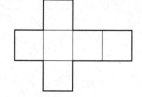

For example, the diagram shows a net of a cube.

To make the cube, the net is folded as shown below.

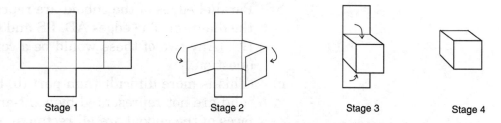

Stage 1　　　Stage 2　　　Stage 3　　　Stage 4

A solid can have several different nets. The net of a cube must consist of six squares (they become the faces of the cube) but they may be arranged in different ways.

Here are two more nets for a cube.
How many more can you think of?

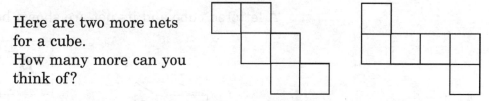

In practice, to make the cube rigid, you will need glue flaps on some of the edges, but these are not usually shown on the net of the solid.

Example 1

cone   cube   cuboid   cylinder   pyramid   sphere   triangular prism

From the list of names above, choose the best one to describe each of the following solids.

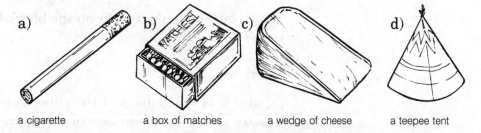

a) a cigarette　　b) a box of matches　　c) a wedge of cheese　　d) a teepee tent

Solution

a) cylinder
b) cuboid
c) triangular prism
d) cone

Example 2

The diagram represents a cuboid.
a) How many edges does the cuboid have?
b) Name two edges of the cuboid which are parallel to the edge BC.
c) Name two edges of the cuboid which are perpendicular to the edge BC.

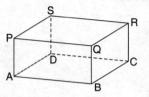

Solution

a) The cuboid has 12 edges (4 at the base, 4 at the top, 4 'vertical').
b) Parallel edges of the cuboid are represented by parallel lines in the diagram. The edges AD, PS and QR are parallel to the edge BC. (Any two of these would be a correct answer to the question.)
c) (This is more difficult than part (b) because right-angles in the solid are not represented by right-angles in the diagram.) The faces of the cuboid are all rectangular. It follows that the edges BA, BQ, CD and CR are perpendicular to the edge BC. (Any two of these would be a correct answer to the question.)

Example 3

A lettered cube and its net are shown below.

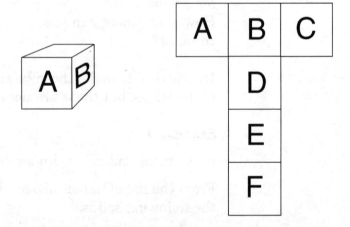

Which letter should appear on the blank face shown on the top of the cube?

Solution

Square C is at the back of the cube (opposite to square A).
Square D is the bottom face of the cube.
Square E is the left-hand face (opposite to square B).
So square F is on the top of the cube.

In fact, the cube will look like this.

Example 4

The diagram shows the net of a solid.

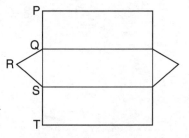

a) What type of solid is this?
b) How many faces has the solid?
c) How many edges has the solid?
d) In an accurately drawn net, which of the lengths PQ, QR, RS, ST *must* be equal?

Solution

a) The solid is a triangular prism. This diagram shows its shape.

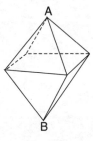

b) The solid has 5 faces (the same as the number of regions in the net).
c) The solid has 9 edges.
d) PQ must be equal in length to QR. (They come together to form one edge of the solid.) Similarly, RS must be equal in length to ST.

Example 5

This diagram represents a solid with 8 faces. Each face is an equilateral triangle. (The solid is a regular octahedron.)

a) How many edges does the solid have?
b) How many vertices does the solid have?
c) Draw a net of the solid.

Solution

a) The octahedron has 12 edges (4 meet at A, 4 meet at B and the other 4 form a square round the 'middle').
b) The octahedron has 6 vertices (A, B and the vertices of the square).
c) The diagram shows one possible net of the octahedron. It consists of 8 identical equilateral triangles. The 4 triangles with A as vertex become the upper half of the solid (this is a square based pyramid) and the 4 triangles with B as vertex become the lower half of the solid.

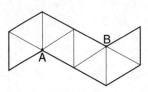

It is time for you to try some questions for yourself. If you have any difficulty, you should examine an everyday object which has the shape mentioned in the question, or make a model from a net.

EXERCISE 19

1. cone   cube   cuboid   cylinder   prism   pyramid   sphere

   From the list of names above, choose the best one to describe each of the following solids.

   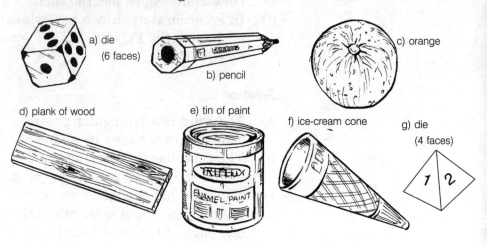

   a) die (6 faces)   b) pencil   c) orange
   d) plank of wood   e) tin of paint   f) ice-cream cone   g) die (4 faces)

2. The diagram represents a view of an ordinary die. The number of dots on opposite faces add up to seven.

   On this net of the die, put the correct number of dots on the blank faces.

   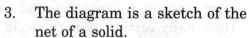

3. The diagram is a sketch of the net of a solid.
   a) What type of solid is this?
   b) When the net is drawn accurately,
      (i) why must PB and PA each be 50 mm long?
      (ii) what special type of triangle will ADS be?
      (iii) what will be the size of the angle marked x?
   c) How many faces, edges and vertices does the solid have?

   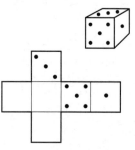

4. The diagram represents a small box. Each of the faces is a rectangle.
   a) Give the mathematical name for the shape of the box.
   b) How many faces, edges and vertices does the box have?
   c) Draw an accurate net for the box.
   d) How many of these boxes would fit into a carton 8 cm by 9 cm by 11 cm?

   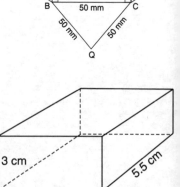

5. A cuboid is 3 cm long, 3 cm wide and 2 cm high.
   Its surface is painted red and then the cuboid is cut up into 1 cm cubes, as shown in the diagram.
   a) How many 1 cm cubes are there?
   b) Write down the number of 1 cm cubes which have:
      (i) 3 faces painted red
      (ii) exactly 2 faces painted red
      (iii) only 1 face painted red

You may have found this work on 3-dimensional objects more of a challenge than the 2-dimensional work you did earlier in Module 4. Check your answers at the end of this module.

Learners following the CORE syllabus should now move on to Section B. The work on symmetry of solids is for learners following the EXTENDED syllabus.

Symmetry of solids

In Unit 1 Section C you learnt about the symmetry of 2-dimensional shapes. You will remember that this is of two types – line symmetry and rotational symmetry.

The symmetry of 3-dimensional shapes – that is, solids – is also of two types. These are called plane symmetry (also known as 'reflective symmetry') and rotational symmetry.

Plane symmetry

A solid has plane symmetry if it can be cut into two halves and each half is the mirror image of the other half. The plane separating the two halves is called the **plane of symmetry**.

A solid can have more than one plane of symmetry.

Here are some examples. (In each diagram, a plane of symmetry is shaded.)

A cuboid (length, width and height all different) has 3 planes of symmetry.

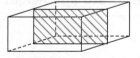 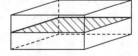 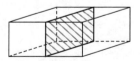

A right, square based pyramid has 4 planes of symmetry.

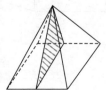 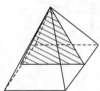 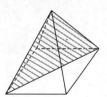

A prism with isosceles triangular ends has 2 planes of symmetry.

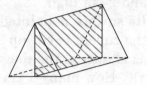 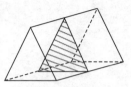

A cube has 9 planes of symmetry.

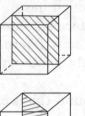

A sphere has an infinite number of planes of symmetry. It is symmetrical about any plane which passes through its centre.

Rotational symmetry

A solid has rotational symmetry if it can be rotated about a line through an angle less than 360° and still look the same. The line about which the solid is rotated is called an **axis of symmetry**.

A solid can have more than one axis of symmetry.

Here are examples. (In each diagram, one axis of symmetry is shown.)

A cuboid (length, width and height all different) has 3 axes of symmetry.

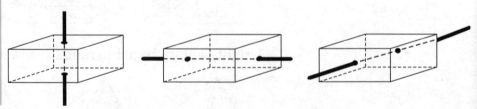

A right, square based pyramid has 1 axis of symmetry.

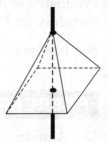

A prism with isosceles triangular ends has 1 axis of symmetry.

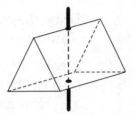

A cube has 13 axes of symmetry.

3 join the centres of opposite faces of the cube.

 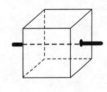

6 join the mid-points of opposite edges of the cube.

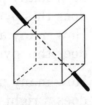 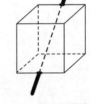 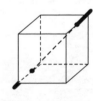

 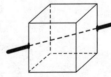

4 join opposite vertices of the cube.

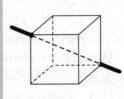 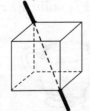 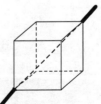

A sphere has an infinite number of axes of symmetry. It has rotational symmetry about every diameter.

You will probably find it difficult to identify the planes and axes of symmetry of a solid by looking at 2-dimensional sketches. You will certainly find it easier if you have the actual solid in your hands.

See how you get on with the questions in this exercise.

EXERCISE 20

1. A prism with equilateral triangular ends has 4 planes of symmetry and 4 axes of rotational symmetry.

   Draw diagrams to show these planes and axes.

   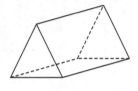

2. A cuboid has dimensions 3 cm by 3 cm by 5 cm.
   a) How many planes of symmetry does it have?
   b) How many axes of symmetry does it have?

   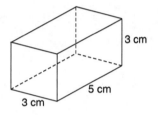

3. A pyramid VABCD has a rectangular base ABCD.
   AB = 5 cm, BC = 3 cm and VA = VB = VC = 7 cm
   a) How many planes of symmetry does this pyramid have?
   b) How many axes of symmetry does this pyramid have?

   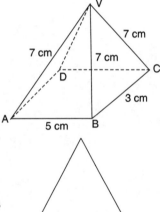

4. a) How many planes of symmetry does a right circular cone have?
   b) How many axes of symmetry does a right circular cone have?

5. This solid 'plus' sign is made of 5 cubes, all the same size.

   How many planes of symmetry does this solid have?

6. 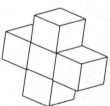 This diagram represents a tea cup.
   a) How many planes of symmetry does the cup have?
   b) How many axes of symmetry does the cup have?

Check your answers at the end of this module.

Did you find this exercise difficult? If so, don't worry – symmetry of solids is a relatively minor item in the IGCSE EXTENDED syllabus. If most of your answers are correct, congratulations!

B Congruent figures and similar figures

In Unit 2 Section A, when you were learning about drawing triangles, I introduced the idea of congruent triangles and similar triangles. In fact, the idea of congruence and similarity can be extended to other plane figures, and to solids also.

Figures or solids which are exactly the same shape are said to be **similar**.

Figures or solids which are exactly the same shape *and* the same size are said to be **congruent**. It may be necessary to rotate the figures or solids, or to reflect one of them in a mirror, in order to determine whether they are similar or congruent.

For the IGCSE examinations, you have to be able to recognise similar figures and congruent figures. You need not worry about similar and congruent solids.

In simple cases, you should be able to decide whether two figures are similar, congruent, or neither, just by examining them carefully. When testing for congruency, it may be worthwhile to trace one figure and then place the tracing over the other figure.

For similarity, the angles in one figure must be exactly the same, and in the same order, as in the other figure, *and* corresponding sides must be in the same ratio. (One figure can be regarded as an enlargement of the other figure. The ratio of corresponding sides is the scale of the enlargement.)

Example 1

In each of the following, state which of the three figures is not congruent to the other two.

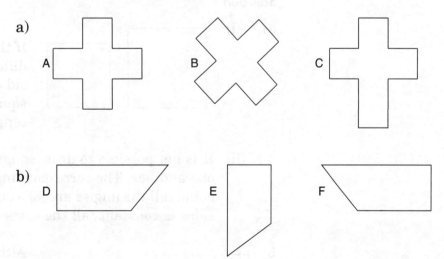

Solution

a) A and B are congruent to one another. [B is A rotated through 45°]

   C is not congruent to A and B. [one 'leg' is longer that the others]

b) D and F are congruent. [they are mirror images of one another]

   E is not congruent to D or F. [the parallel sides are 2.5 cm and 1.5 cm instead of 3 cm and 2 cm]

Example 2

In each of the following, state which of the figures is not similar to the other two.

a)

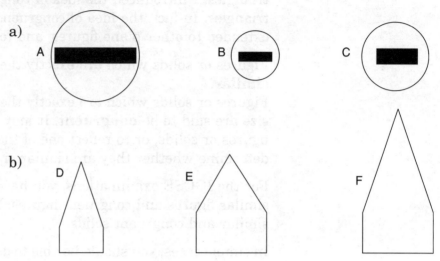

b)

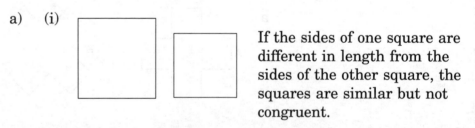

Solution

a) A is not similar to the other two. [the 'minus sign' is too long]

b) E is not similar to the other two. [the angle at the top of E is larger than the angle at the top of D or F]

Example 3

a) (i) Draw two squares which are not congruent to one another.
   (ii) Is it possible to draw two squares which are not similar to one another?
b) Is it possible to draw rectangles which are not similar to one another?

Solution

a) (i)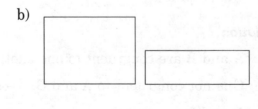

If the sides of one square are different in length from the sides of the other square, the squares are similar but not congruent.

(ii) It is not possible to draw squares which are not similar to one another. The corresponding angles of the squares are equal (all the angles are 90°) and the ratio of corresponding sides is constant (all the sides of a square are equal).

b) Although the corresponding angles of these rectangles are equal (all the angles are 90°), the ratio of corresponding sides is not constant. The rectangles are different shapes so they are not similar.

Example 4

a) A regular hexagon is divided into four triangles, as shown in the diagram. Which of the triangles are congruent to one another?

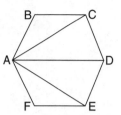

b) Show how a regular hexagon can be divided into four triangles, three of which are congruent to one another.

Solution

a) The figure is symmetrical about AD so triangle ABC is congruent to triangle AFE, and triangle ACD is congruent to triangle AED.

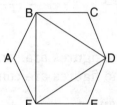

Triangles ABF, BCD and DEF are congruent to one another. (The figure has rotational symmetry about the centre of the hexagon.)

Example 5

Peta has a photograph which she wants to enlarge.
The photograph is a rectangle with height 8 cm and width 6 cm.
The enlargement will have a height of 12 cm.
What will be the width of the enlargement?

Solution

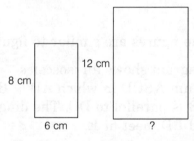

Since one rectangle is an enlargement of the other, the two rectangles are similar.

The ratio of the heights $= \frac{12}{8} = 1.5$ and so the ratio of the widths must be 1.5.

Hence, the width of the enlargement $= 1.5 \times 6$ cm $= 9$ cm.

Did you find it easy to follow this work on similar and congruent figures? If so, you should have no difficulty in answering the following questions.

EXERCISE 21

1. Look at these figures.

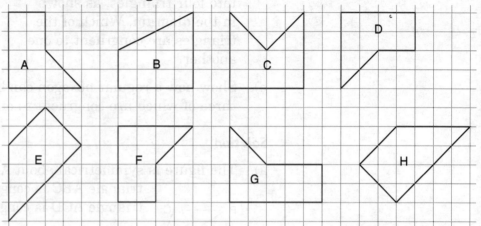

   a) Which of the figures are congruent to figure A?
   b) Which of the figures are congruent to figure E?

2. Look at these figures.

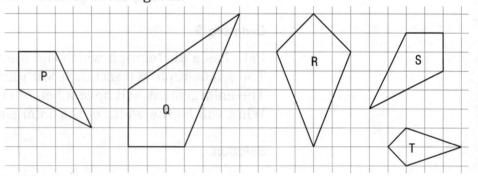

   Which of the figures are similar to figure P?

3. The diagram shows an isosceles trapezium ABCD in which AD = BC and AB is parallel to DC. The diagonals AC and BD meet at E.
   a) Name two triangles in the diagram which are congruent.
   b) Name two triangles in the diagram which are similar but not congruent.

   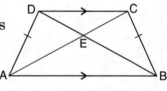

4. The diagram shows an isosceles trapezium ABCD.

   Three trapeziums, congruent to ABCD, fit together to form an equilateral triangle PQR with a side of 4.5 cm.

   Draw the three trapeziums in this diagram.

   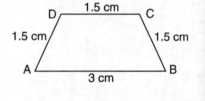

   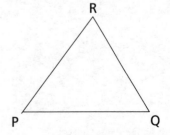

5. 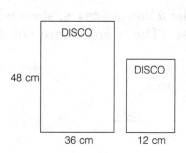 To advertise a disco, the organisers are printing posters and handbills. The posters are an enlargement of the handbills. Each poster is a rectangle with length 48 cm and width 36 cm. Each handbill has a width of 12 cm.
   a) Calculate the length of each handbill.
   b) The letters 'DISCO' on the handbills are 9 mm high. What is the height of these letters on the posters?

Check your answers at the end of this module.

Similar triangles

You have already seen that, for two figures to be similar, they must have their corresponding angles equal *and* their corresponding sides in the same ratio.

It is possible for two figures to satisfy one of these conditions but not the other, and, in this case, the figures are *not* similar.

These trapeziums have corresponding angles equal but the corresponding sides are not in the same ratio. The trapeziums are *not* similar.

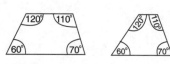

These quadrilaterals have corresponding sides in the same ratio (2:1) but their corresponding angles are not equal. The quadrilaterals are *not* similar.

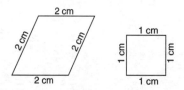

However, there is a surprise in store when you consider triangles. If two triangles satisfy one of the conditions, then they automatically satisfy the other condition, and so they are similar.

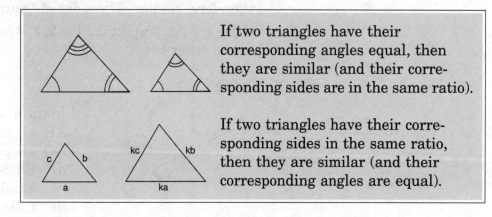

If two triangles have their corresponding angles equal, then they are similar (and their corresponding sides are in the same ratio).

If two triangles have their corresponding sides in the same ratio, then they are similar (and their corresponding angles are equal).

Example 1

In each of the following cases, state which triangle is not similar to the other two. (The triangles are not drawn to scale.)

a)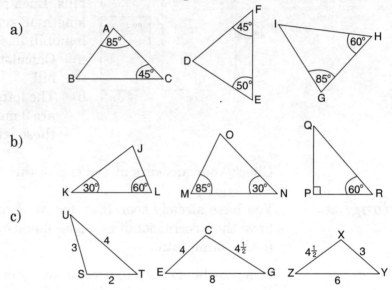

b)

c)

Solution

a) Using angle sum of a triangle = 180°, we obtain B = 50°, D = 85° and I = 35°.
    Triangles ABC and DEF have the same angles (45°, 50°, 85°) so they are similar. The angles of triangle GHI are 35°, 60°, 85° so this triangle is not similar to the other two.

b) The angles of triangles JKL and PQR are 30°, 60°, 90° so they are similar. The angles of triangle MNO are 30°, 65°, 85° so this triangle is not similar to the other two.

c) Put in numerical order, the sides of triangle STU are 2, 3, 4 and the sides of the triangle XYZ are 3, $4\frac{1}{2}$, 6. Hence the corresponding sides are in the ratio 2:3 and the triangles are, therefore, similar. The sides of triangle CEG are 4, $4\frac{1}{2}$, 8 and these are not proportional to 2, 3, 4 so this triangle is not similar to the other two.

Example 2

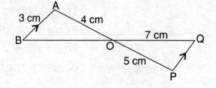

In the diagram, BA is parallel to PQ, AB = 3 cm, AO = 4 cm, OP = 5 cm, OQ = 7 cm.

a) Show that triangle ABO is similar to triangle PQO.
b) Calculate the lengths of BO and PQ.

> It is useful to name the vertices of the triangles so that the corresponding angles (and corresponding sides) are easily identified. In this example, the order ABO corresponds to the order PQO because angle A = angle P, angle B = angle Q and angle O (in triangle ABO) = angle O (in triangle PQO).

Solution

a) In triangles ABO and PQO,
angle BAO = angle QPO (alternate angles, BA parallel to PQ)
angle ABO = angle PQO (alternate angles, BA parallel to PQ)
angle AOB = angle POQ (vertically opposite angles)

Hence, triangle ABO is similar to triangle PQO (corresponding angles equal)

b) In these triangles, side AB corresponds to side PQ,
    side BO corresponds to side QO,
    side OA corresponds to side OP.

The triangles are similar so the corresponding sides are in the same ratio.

That is $\frac{AB}{PQ} = \frac{BO}{QO} = \frac{OA}{OP}$ so $\frac{3}{PQ} = \frac{BO}{7} = \frac{4}{5}$

Hence, $BO = \frac{4 \times 7}{5} = 5.6$ cm and $PQ = \frac{3 \times 5}{4} = 3.75$ cm.

Example 3

The diagram shows a street lamp AB which is 6.8 m high. A girl, who is 1.7 m tall, stands 7.5 m away from the point B. Her shadow is x metres long.

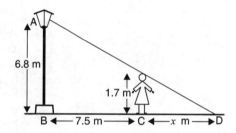

a) Explain why $\frac{x}{x+7.5} = \frac{1.7}{6.8}$.

b) Find the value of x.

Solution

In the diagram, CE represents the girl. AB and EC are vertical so they are parallel.

a) In the overlapping triangles ABD and ECD,
angle BAD = angle CED   (corresponding angles
angle ABD = angle ECD   AB parallel to EC)
angle ADB = angle EDC   (they are the same angle)

Hence the triangles are similar and

$\frac{CD}{BD} = \frac{CE}{BA}$

That is $\frac{x}{x+7.5} = \frac{1.7}{6.8}$

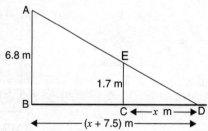

b) The equation is equivalent to $\frac{x}{x+7.5} = \frac{1}{4}$

Hence $4x = x + 7.5$
and so $3x = 7.5$
The value of x is 2.5.

If you are following the CORE syllabus you can now do questions 3 and 6 of Exercise 22 and then move on to the 'Summary' – you have covered all the parts of Module 4 which are in your syllabus. If you are following the EXTENDED syllabus you must do some further work on congruency and circles.

Congruent triangles

If two triangles are congruent, they have their corresponding angles equal and their corresponding sides equal.

However, as you found when you were drawing triangles in Unit 2 Section A, it is not necessary to know the sizes of all the angles and the lengths of all the sides of a triangle to fix its shape and size. For example, if two people were asked to draw a triangle with sides of length 3 cm, 4 cm, 5 cm, their triangles will be exactly the same shape and size. This means that, if the three sides of one triangle are equal to the three sides of another triangle, you can be sure that the triangles are congruent.

As you know, there are other cases where three pieces of information about a triangle is sufficient to fix the shape and size of the triangle. These cases give us sets of conditions for two triangles to be congruent.

> If three sides of one triangle are equal to the three sides of another triangle, then the two triangles are congruent.

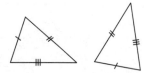

We refer to this set of conditions as '3 sides' or 'side, side, side' or 'SSS'.

> If two sides of one triangle are equal to two sides of another triangle and the angle between this pair of sides is the same in both triangles, then the two triangles are congruent.

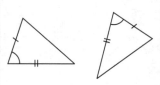

We refer to this set of conditions as '2 sides and the included angle' or 'side, angle, side' or 'SAS'.

> If two angles of one triangle are equal to two angles of another triangle and the side between this pair of angles is the same length in both triangles, then the two triangles are congruent.

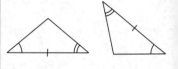

We refer to this set of conditions as 'angle, side, angle' or 'ASA'.

If two angles of one triangle are equal to two angles of another triangle, the third angle in the first triangle must be equal to the third angle in the other triangle. (The sum of the three angles of a triangle is always 180°.) You may have to use this fact before you can apply the ASA condition.

| The longest side of a right-angled triangle is always opposite the right-angle. This side is known as the **hypotenuse** of the triangle. |

> If two right-angled triangles have their longest sides equal in length and another side of the first triangle is equal to a side of the other triangle, then the two triangles are congruent.

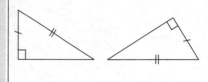

We refer to this set of conditions as 'right-angle, hypotenuse, side' or 'RHS'.

Example 1

In each of the following cases, state whether the two triangles are congruent and, if they are, give a reason. (The diagrams are not drawn to scale.)

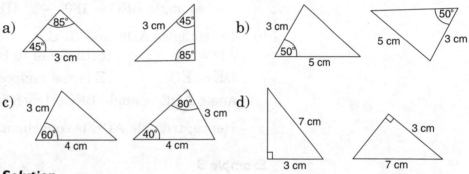

Solution

a) In each triangle, the third angle = 180° − (85° + 45°) = 50°.
   In each triangle, the 3 cm side has angles of 45° and 50° at its ends. Hence, the triangles are congruent (ASA).

b) In the first triangle, the angle of 50° is between the 3 cm and 5 cm sides (it is the 'included angle').
   In the second triangle, the angle of 50° is not between the 3 cm and 5 cm sides.
   We cannot be sure whether the triangles are congruent. (If they are, each of them would have two angles of 50° and two sides of 5 cm.)

c) In the second triangle, the third angle = 180° − (80° + 40°) = 60°.
   Both triangles have sides of 3 cm and 4 cm with an included angle of 60°. Hence, the triangles are congruent (SAS).

d) Both of these right-angled triangles have a hypotenuse of 7 cm and one side of 3 cm. Hence, the triangles are congruent (RHS).

Example 2

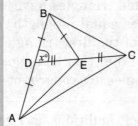

In this triangle ABC,
D is the midpoint of AB,
E is the midpoint of CD,
BE = BD and angle BDE = $x°$.
a) Find, in terms of x,
   (i) angle ADE   (ii) angle BEC
b) Name a triangle which is congruent to triangle BEC.

Solution

a) (i) angle ADE = $180° - x°$ (BDA is a straight angle)
   (ii) angle BED = $x°$ (base angles of isosceles triangle BDE)
       so angle BEC = $180° - x°$ (DEC is a straight angle)

b) In triangles ADE and BEC,
   AD = BE   (both equal to BD)
   DE = EC   (E is the midpoint of CD)
   angle ADE = angle BEC   (each is $180° - x°$)

   Hence, triangle ADE is congruent to triangle BEC. (SAS)

Example 3

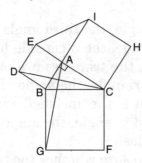

The diagram shows triangle ABC, which is right-angled at A, with squares ABDE, BCFG, ACHI drawn on its sides.
a) E is joined to I.
   Prove that triangle ABC is congruent to triangle AEI.
b) A is joined to G and C is joined to D.
   Prove that triangle ABG is congruent to triangle DBC.

Solution

a) In triangles ABC and AEI,

   AB = AE   (sides of square ABDE)
   AC = AI   (sides of square ACHI)
   angle BAC = angle EAI   (vertically opposite angles)
   Hence, triangle ABC is congruent to triangle AEI. (SAS)

b) angle ABG = 90° + angle ABC   (ABFG is a square so
                                     angle CBG = 90°)

   angle DBC = 90° + angle ABC   (ABDE is a square so
                                     angle ABD = 90°)

   It follows that angle ABG = angle DBC.

In triangles ABG and DBC,
angle ABG = angle DBC (proved)
AB = DB (sides of square ABDE)
BG = BC (sides of square BCFG)

Hence, triangle ABG is congruent to triangle DBC. (SAS)

Note: From these two results, you could make various deductions.

For example, BC = EI (triangles ABC and AEI are congruent)
AG = DC (triangles ABG and DBC are congruent)
angle ABC = angle AEI (triangles ABC and AEI are congruent)

Here are some questions on similar triangles and congruent triangles for you to try.

EXERCISE 22

1. In each of the following cases, state which of the three triangles is not similar to the other two. (Triangles are not drawn to scale.)

   a)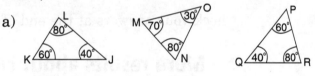

   b)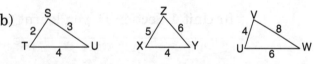

   c)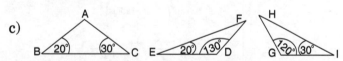

2. In each of the following cases, state whether the two triangles are congruent and, if they are, give a reason.

   a)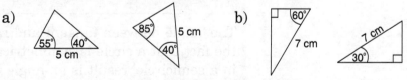

   b)

   c) 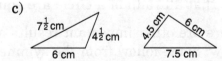 d)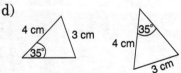

3. In the diagram, SR is parallel to PQ, SR = 4 cm, SX = 3 cm, RX = 2 cm and PQ = 7 cm.
   a) Explain why triangle RSX is similar to triangle PQX.
   b) Calculate the length of PX and the length of QX.

   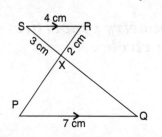

4. In this diagram, O is the centre of the circle and AB is a chord of the circle. ON is the perpendicular from O to AB.

   Prove that triangle OAN is congruent to triangle OBN.

5. ABC and APQ are equilateral triangles with a common vertex A.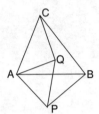
   a) Explain why angle CAQ = angle BAP.
   b) Prove that triangle CAQ is congruent to triangle BAP.

6. 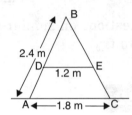 The diagram represents a step ladder made of two sections, AB and BC. DE represents a bar which is used to stabilise the ladder. DE is parallel to AC. AB = 2.4 m, DE = 1.2 m and AC = 1.8 m.
   a) Explain why triangles BDE and BAC are similar.
   b) Calculate the length of BD.
   c) Given that EC = 0.85 m, calculate the length of BE.

Check your answers at the end of this module.

C More results about circles

In Unit 1 Section D, you learnt two important results about circles.

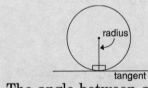

The angle between a tangent and the radius through the point of contact is 90°.

The angle in a semicircle is 90°.

The 'angle between tangent and radius' result is a consequence of the fact that a circle is symmetrical about any diameter. The 'angle in a semicircle' result is an angle property which follows from the fact that all radii of a circle are equal in length.

There are other important results about circles which you need to know. Some follow from the symmetries of a circle and some are angle properties which are not directly related to symmetry.

Symmetry properties of a circle

A circle has line symmetry about any diameter and it has rotational symmetry about its centre. From these facts, we can deduce a number of results.

> The perpendicular bisector of a chord passes through the centre.

The perpendicular bisector of the chord AB is the locus of points which are equidistant from A and B. But the centre, O, of the circle is equidistant from A and B (because OA and OB are radii of the circle). Hence, O must be on the perpendicular bisector of the chord AB. (This perpendicular bisector is the line of symmetry of the line segment AB and a line of symmetry of the circle.)

This result can be expressed in other ways:
a) The perpendicular from the centre of a circle to a chord meets the chord at its midpoint.
b) The line joining the centre of a circle to the midpoint of a chord is perpendicular to the chord.

> Equal chords are equidistant from the centre and chords equidistant from the centre are equal in length.

> The distance from a point to a line is always the *perpendicular distance*, which is the shortest distance from the point to the line.

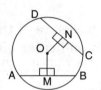

If chords AB and CD are the same length, then OM = ON, and vice versa.

This follows from the fact that triangle OAM is congruent to triangle ODN.

> Can you prove this?

It is also a consequence of the fact that the circle has rotational symmetry about O.

> The two tangents drawn to a circle from a point outside the circle are equal in length.

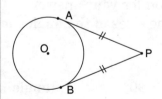

A and B are the points of contact of the tangents drawn from P.

The result is PA = PB.

This is a consequence of the fact that the figure is symmetrical about the line OP. (It can also be obtained by proving that triangle OAP is congruent to triangle OBP. You need to use the 'tangent perpendicular to radius' property for this.)

> The line joining the centre of a circle to a point P outside the circle bisects the angle between the tangents drawn from P.

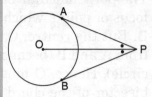

A and B are the points of contact of the tangents drawn from P. The result is angle OPA = angle OPB.

As for the previous result, this is a consequence of the fact that the figure is symmetrical about the line OP (or that triangles OAP and OBP are congruent).

Example 1

The centre of this circle has not been given. Describe how to find the centre.

Solution

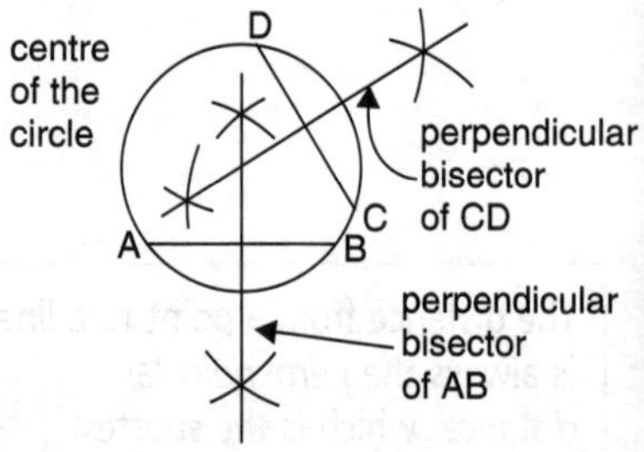

Draw two chords, AB and CD, which are not parallel.

Draw the perpendicular bisectors of these chords. Both of these perpendicular bisectors pass through the centre of the circle. Their point of intersection must be the centre of the circle.

Example 2

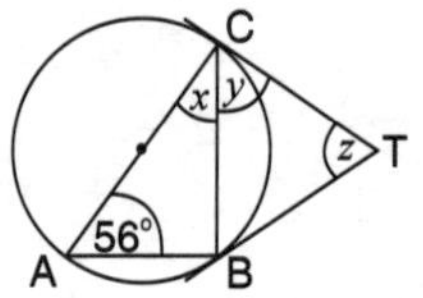

In the diagram, AC is a diameter of the circle and B is a point on the circle.

The tangents to the circle at B and C meet at T, and angle BAC = 56°.
a) What is the size of angle ABC? Give a reason for your answer.
b) Calculate the sizes of the angles marked x, y, z in the diagram.

Solution

a) angle ABC = 90°   (angle in a semicircle)
    $x = 180° - (90° + 56°)$ (angle sum of triangle ABC)
    so $x = 34°$
    angle ACT = 90°   (angle between tangent and radius)
    so $x + y = 90°$
    $y = 90° - 34°$
    $y = 56°$
    CT = BT   (tangents from T)
    so triangle BCT is isosceles
    angle CBT = y (base angles of isosceles triangle BCT)
            = 56°
    so $z = 180° - (56° + 56°)$ (angle sum of triangle BCT)
    $z = 68°$

Example 3

A straight line cuts two concentric circles at A, B, C, D (in that order).

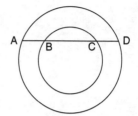

Prove that AB = CD.

Concentric circles means circles which have the same centre.

Solution

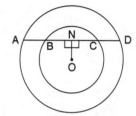

Draw the perpendicular from the centre, O, of the circle to the line ABCD. Label the foot of the perpendicular N.

AN = ND   (the perpendicular from centre bisects chord AD)

BN = NC   (the perpendicular from centre bisects chord BC)

Hence, AN − NB = ND − NC
    That is AB = CD

Example 4

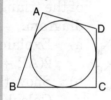

Each of the four sides of a quadrilateral ABCD is a tangent to a circle (see diagram). Prove that AB + CD = AD + BC.

Solution

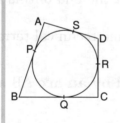

Let the points of contact of the tangents AB, BC, CD, DA be P, Q, R, S, respectively.

Then AP = AS   (tangents from point A)

PB = BQ   (tangents from point B)

CR = QC   (tangents from point C)

RD = SD   (tangents from point D)

Hence,
(AP + PB) + (CR + RD) = (AS + SD) + (BQ + QC)
    That is AB + CD = AD + BC

Using the facts of circles which you have met so far, you should be able to solve the problems in Exercise 23. You will see that if you don't know the facts you will find it difficult to solve the problems. So make sure you understand all the facts and remember them well.

EXERCISE 23

1. P is a point inside a circle whose centre is O.

   Describe how to construct the chord which has P as its midpoint.

2.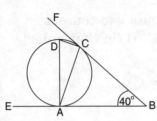

In the diagram, BCF and BAE are the tangents to the circle at C and A respectively. AD is a diameter and angle ABC = 40°.
a) Explain why triangle ABC is isosceles.
b) Calculate the size of:
(i) angle CAB
(ii) angle DAC
(iii) angle ADC

3.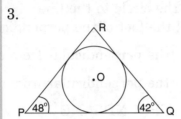

The diagram shows a circle, centre O, which touches the sides of the triangle PQR.

Given that angle RPQ = 48° and angle PQR = 42°, calculate the size of angle POQ.

4.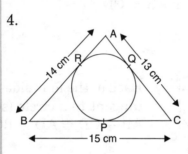

In triangle ABC, AB = 14 cm, BC = 15 cm and CA = 13 cm. A circle touches the sides of the triangle at P, Q, R, as shown in the diagram.
a) Explain why AB + BC + CA = 2(AR + BP + PC).
b) Show that AR = 6 cm.
c) Calculate the length of BP.

Check your answers at the end of this module.

More angle properties of a circle

In these properties some technical terms are used which I need to explain.

'The angle **subtended by an arc** AB **at the centre**' means the angle AOB.

If AB is a minor arc, angle AOB is less than 180°.

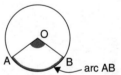

If AB is a semicircular arc, angle AOB is 180°.

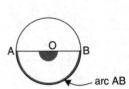

If AB is a major arc, angle AOB is more than 180° (it is a reflex angle).

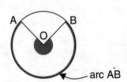

'The angle **subtended by an arc** AB **at the circumference**' means the angle APB where P is a point on the circumference excluding the arc AB.

If AB is a minor arc, angle APB is acute.

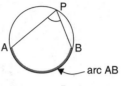

If AB is a semicircular arc, angle APB is 90°.

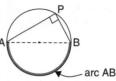

If AB is a major arc, angle APB is obtuse.

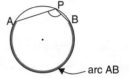

'Angles in the **same segment**' means two or more angles (APB, AQB in these diagrams) subtended by the *same* arc (AB) at the circumference.

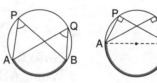 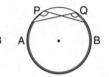

'A **cyclic quadrilateral** ABCD' means a quadrilateral which has all four of its vertices A, B, C, D on the circumference of the same circle.

Now I'll state the results you need to know. The 'angle' at the centre theorem is the main result. The other results are simple deductions from that theorem.

> The angle subtended at the centre of a circle by an arc is twice the angle subtended at the circumference by the arc.

 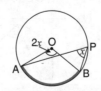

If AB is an arc of a circle with centre O, and P is a point on the circumference but not on the arc AB, the angle at the centre theorem states angle AOB = 2 × angle APB.

The theorem can be proved as follows:

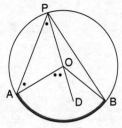

Join P to O and extend this line to D.

OP = OA (radii) so triangle OAP is isosceles.

angle OAP = angle OPA (base angles of isosceles triangle)

angle AOD = angle OAP + angle OPA (exterior angle of triangle OAP)

Hence angle AOD = 2 × angle OPA (1)
Similarly angle BOD = 2 × angle OPB (2)
Adding (1) and (2) gives
angle AOB = 2 × angle APB.

Note 1

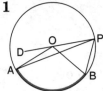

For this diagram, statements (1) and (2) must be *subtracted* to obtain the result angle AOB = 2 × angle APB.

make sure you understand this by doing the proof yourself, using this diagram

Note 2

When AB is a semicircular arc, the angle at the centre theorem becomes the 'angle in a semicircle is 90°'. This is because, in this case, angle AOB is a straight angle (180°).

> Angles in the same segment, subtended by the same arc, are equal.

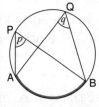 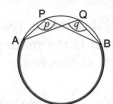

This result states that, in these diagrams, $p = q$.

The result is true because each of the angles p and q is half the angle subtended by the arc AB at the centre of the circle.

> Opposite angles of a cyclic quadrilateral add up to 180°.

$a + c = 180°$
and
$b + d = 180°$

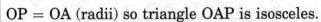

In this diagram,

$\alpha = 2a$ (angle at centre theorem using minor arc BD)

$\beta = 2c$ (angle at centre theorem using major arc BD)

Hence, $\alpha + \beta = 2(a + c)$

But $\alpha + \beta = 360°$ (one complete turn), so $a + c = 180°$.
The angle sum of a quadrilateral is 360°, so $b + d = 360° - (a + c) = 180°$.

Each exterior angle of a cyclic quadrilateral is equal to the interior angle opposite to it.

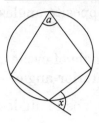

In this diagram, $x = a$, and there are similar results for the other three exterior angles.

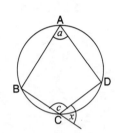

Here is the proof:
In the diagram,
$a + c = 180°$ (opposite angles of a cyclic quadrilateral)

but $x + c = 180°$ (a straight angle)

Hence, $x = a$

Example 1

Find the size of each lettered angle in these sketches. When it is marked, O is the centre of the circle.

a) 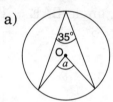   b) 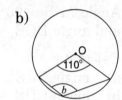   c)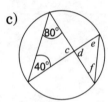

Solution

a) $a = 2 \times 35°$ (angle at the centre theorem)
   $a = 70°$

b) reflex angle at O $= 360° - 110° = 250°$
   $b = $ (reflex angle at O) $\div 2 = 250° \div 2$
   $b = 125°$

c) $c = 180° - (40° + 80°)$ (angle sum of triangle)
   $c = 60°$
   $d = 60°$ (vertically opposite to angle c)
   $e = 80°$ (angles in same segment)
   $f = 40°$ (angles in same segment)

Example 2

Find the size of each lettered angle in these sketches.

a) 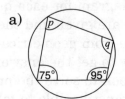   b) 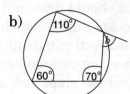   c)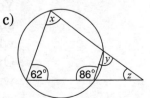

Solution

a) $p + 95° = 180°$ (opposite angles of cyclic quadrilateral)
    $p = 85°$
    $q + 75° = 180°$ (opposite angles of cyclic quadrilateral)
    $q = 105°$

b) $b = 60°$ (exterior angle of cyclic quadrilateral equals the opposite interior angle)

c) $x + 86° = 180°$ (opposite angles of cyclic quadrilateral)
    $x = 94°$
    $y = 62°$ (exterior angle of cyclic quadrilateral)
    $x + z + 62° = 180°$ (angle sum of a triangle)
    $z = 180° - (94° + 62°)$
    $z = 24°$ you could also use the fact that $y + z = 86°$ (exterior angle of triangle) to get the same answer for z

Example 3

In the diagram, the points B, C, D, E lie on the circle.
ABC and AED are straight lines.
Angle AEB = 96°, angle BAE = 19° and angle DCE = 78°.

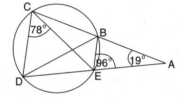

Calculate:
a) angle BCE
b) angle CDE
c) angle CDB

Solution

a) angle BCD = 96°   (exterior angle of cyclic quadrilateral)
    Hence, angle BCE = 96° − 78°
    angle BCE = 18°

b) angle ABE = 180° − (96° + 19°)   (angle sum of triangle ABE)
            = 65°
    angle ABE = angle CDE (exterior angle of cyclic quadrilateral)
    angle CDE = 65°

c) angle BDE = angle BCE (angles in the same segment)
            = 18°
    angle CDB = angle CDE − angle BDE = 65° − 18°
    angle CDB = 47°

Here are some angle calculations for you to do.

You should draw a diagram for each question, and write on it the sizes of angles which are given and the sizes of any other angles which you can deduce from geometrical results which you know. Once you can see how to get the answer, work out in which order to write the steps you took to solve the problem. Remember that if I read your solution I must be able to follow it clearly step by step. When you write out your solutions, remember to state the geometrical results which you have used at each step.

EXERCISE 24

1. Find the size of each lettered angle in these sketches. When it is marked, O is the centre of the circle.

   a) 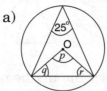 b) 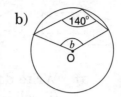 c)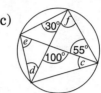

2. Find the size of each lettered angle in these sketches.

   a) 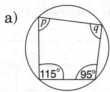 b) 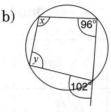 c)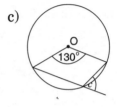

3. In the diagram, SAT is the tangent to the circle at the point A. The points B and C lie on the circle and O is the centre of the circle. If angle ACB = x, express in terms of x, the size of:
   a) angle AOB
   b) angle OAB
   c) angle BAT

   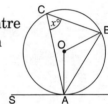

Check your answers at the end of this module.

Summary

In this final unit of Module 4 you learnt about polyhedra:
- how to draw polyhedra
- calculating the faces, vertices and edges
- drawing nets
- plane symmetry and rotational symmetry.

You learnt that for similar figures:
- the angles are equal and
- the corresponding sides are in the same ratio
- to show that two triangles are similar it is sufficient to show that the angles are equal *or* the corresponding sides are in the same ratio.

There are four conditions you can use to prove that two triangles are congruent:
- side, side, side (SSS)
- side, angle, side (SAS)
- angle, side, angle (ASA)
- right angle, hypotenuse, side (RHS)

Check your progress

1. Look at these diagrams of six solids.

   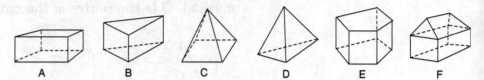

   a) Write down the mathematical name given to:
   (i) solid A   (ii) solid B   (iii) solid C

   b) Complete this table to show the number of faces, the number of vertices and the number of edges of each solid.

Solid	Number of		
	Faces	Vertices	Edges
A			
B			
C			
D			
E			
F			

   c) For these solids, the number of faces + the number of vertices is related to the number of edges. What is the relationship?

2. a) Draw a line to divide this L-shaped diagram into two congruent shapes.

   b) Draw lines to divide this L-shaped diagram into three congruent shapes.

   c) Draw lines to divide this L-shaped diagram into six congruent shapes.

3. If you take a strip of paper with parallel edges, tie a knot in it, and carefully press the knot flat, you will find that a regular pentagon is made. The diagram shows the knot. The hidden edges of the strip of paper are shown as dashed lines. ABCDE is the regular pentagon.

   a) Write down the triangles in the diagram which are congruent to triangle ABX.

   b) Write down the triangles in the diagram which are similar to triangle ABX but not congruent to it.

4. The diagram is the net of a solid.
   a) How many planes of symmetry does the solid have?
   b) How many axes of rotational symmetry does the solid have?

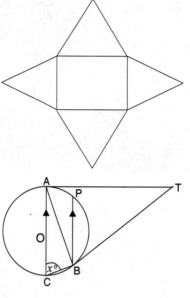

5. In the diagram, TA and TB are the tangents from T to the circle whose centre is O.

   AC is a diameter of the circle and angle ACB = x.
   a) Find angle CAB in terms of x.
   b) Find angle ATB in terms of x.
   c) The point P on the circumference of the circle is such that BP is parallel to CA. Express angle PBT in terms of x.

Check your answers at the end of this module.

Solutions

EXERCISE 1

1. $\frac{5}{6}$ turn $= \frac{5}{6} \times 360° = 5 \times 60° = 300°$
2. $315° = \frac{315}{360}$ turn $= \frac{63}{72}$ turn $= \frac{7}{8}$ turn
3. a) about $\frac{2}{3}$ of $90°$ = about $60°$
   b) $90°$ + about $\frac{1}{2}$ of $90°$ = about $135°$
   c) $180°$ + about $\frac{1}{2}$ of $90°$ = about $225°$
4. a) p and q are acute; n is obtuse
   b) s and t are acute; r and u are obtuse
   c) v and z are acute; w and y are obtuse; x is reflex

5.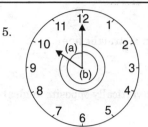
   a) The acute angle $= \frac{2}{3}$ of $90°$ or $\frac{2}{12}$ of a turn
      $= 60°$
   b) The reflex angle $= 360° - 60°$ or $\frac{10}{12}$ of a turn
      $= 300°$

EXERCISE 2

To measure the angles, you need to extend the arms.
Your answers should be within 1 of the values given below.

1. a) $50°$ b) $136°$ c) $110°$
2. a) $360° - 40° = 320°$ b) $360° - 122° = 238°$
3. a) $d = 115°, e = 30°, f = 35°$
   b) $p = 35°, q = 55°, r = 90°$
4. a)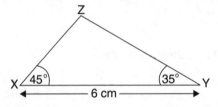
   b) angle $Z = 100°$

5. a)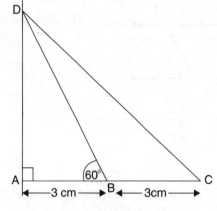
   b) angle BCD = $41°$
   c) CD = 7.9 cm

EXERCISE 3

1. a) $a + 174° + 47° + 50° = 360°$ (angles at a point)
         $a + 271° = 360°$
         so $a = 89°$
   b) $50° + b + 25° = 180°$ (angles on a straight line)
         $b + 75° = 180°$
         so $b = 105°$
   c) $c + 68° = 90°$ (angles in a right-angle)
         so $c = 22°$
2. a) $d + 90° + d + 170° = 360°$ (angles at a point)
         $2d + 260° = 360°$
         $2d = 100°$
         so $d = 50°$
   b) $e + 140° + e = 180°$ (angles on a straight line)
         $2e + 140° = 180°$
         $2e = 40°$
         so $e = 20°$

2. c)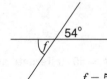

   $f = 54°$ (vertically opposite angles)

   $g + 54° = 180°$ (angles on a straight line)
   so $g = 126°$

   $h + 54° = 180°$ (angles on a straight line)
   or $h = g$ (vertically opposite angles)
   so $h = 126°$

EXERCISE 3 (cont.)

3.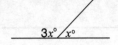

$3x° + x° = 180°$ (angles on a straight line)
$4x° = 180°$
so $x = 45°$

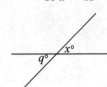

$q° = x°$ (vertically opposite angles)
so $q = 45°$

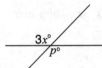

$p° = 3x°$ (vertically opposite angles)
so $p = 135°$

4. a)

$s = 57°$ (alternate angles)

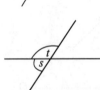

$s + t = 180°$ (angles on a straight line)
so $t = 123°$

b)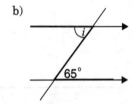

$i = 65°$ (alternate angles)

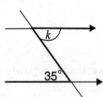

$k = 35°$ (alternate angles)

$i + j + k = 180°$ (angles on a straight line)
$65° + j + 35° = 180°$
$j + 100° = 180°$
so $j = 80°$

4. c)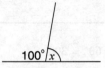

$100° + x = 180°$ (angles on a straight line)
so $x = 80°$

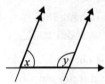

$x + y = 180°$ (co-interior angles)
so $y = 100°$

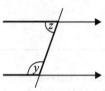

$y + z = 180°$ (co-interior angles)
so $z = 80°$

5.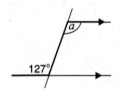

$a = 127°$ (alternate angles)

$a + b = 180°$ (angles on a straight line)
so $b = 53°$

$c = 85°$ (corresponding angles)

Module 4 Solutions

EXERCISE 4

1. a) $x + 65° + 40° = 180°$ (angle sum of a triangle)
    $x + 105° = 180°$
    so $x = 75°$
    $75° + y = 180°$ (angles on a straight line)
    or $y = 65° + 40°$ (exterior angle of a triangle)
    so $y = 105°$

    b)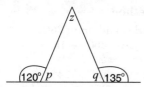

    $p + 120° = 180°$ (angles on a straight line)
    so $p = 60°$
    $q + 135° = 180°$ (angles on a straight line)
    so $q = 45°$

    $p + q + z = 180°$ (angle sum of a triangle)
    $105° + z = 180°$
    so $z = 75°$

    c) $a + 30° + 110° = 180°$ (angle sum of a triangle)
    $a + 140° = 180°$
    so $a = 40°$

    $b + 60° + 80° = 180°$ (angle sum of a triangle)
    $b + 140° = 180°$
    so $b = 40°$

2. a) $a + 110° = 180°$ (angles on a straight line)
    so $a = 70°$
    $b = a$ (base angles of an isosceles triangle)
    so $b = 70°$

    $110° = b + c$ (exterior angle of a triangle)
    or $a + b + c = 180°$ (angle sum of a triangle)
    so $c = 40°$

    b) The two angles marked d are equal and the two angles marked e are equal (base angles of isosceles triangles)
    $d + d + 34° = 180°$ (angle sum of a triangle)
    $2d = 146°$
    so $d = 73°$

    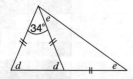

    $d = e + e$ (exterior angle of triangle)
    so $2e = 73°$
    so $e = 36\frac{1}{2}°$

2. c) $f = 78°$ (alternate angles)

    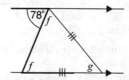

    The two angles marked f are equal (base angles of an isosceles triangle)
    $f + f + g = 180°$ (angle sum of a triangle)
    $156° + g = 180°$
    so $g = 24°$

3. a) $p + 63° + 50° = 180°$ (angle sum of a triangle)
    $p + 113° = 180°$
    so $p = 67°$
    $q + 130° = 180°$ (angles on a straight line)
    so $q = 50°$
    $p + q + r = 180°$ (angle sum of a triangle)
    $117° + r = 180°$
    so $r = 63°$

    b) $x + 64° + 90° = 180°$ (angle sum of a triangle)
    $x + 154° = 180°$
    so $x = 26°$
    $y + 64° + 90° = 180°$ (angle sum of a triangle)
    $y + 154° = 180°$
    so $y = 26°$

    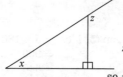

    $z = x + 90°$ (exterior angle of a triangle)
    so $z = 116°$

    c)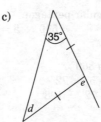

    $d = 35°$ (base angles of isosceles triangle)
    $e = d + 35°$ (exterior angle of a triangle)
    so $e = 70°$

    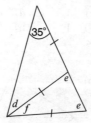

    The two angles marked e are equal (base angles of isosceles triangle)

    $e + e + f = 180°$ (angle sum of a triangle)
    $140° + f = 180°$
    so $f = 40°$

4. a) angle R $+ 84° + 48° = 180°$ (angle sum of triangle)
    angle R $+ 132° = 180°$
    so angle R $= 48°$

    b) Triangle PQR has two angles equal (Q and R), so it is *isosceles*.

EXERCISE 5

1. a) $a + 90° + 110° + 100° = 360°$ (angle sum of quadrilateral)
    $a + 300° = 360°$
    so $a = 60°$

    b) $x + 123° = 180°$ (angles on a straight line)
    so $x = 57°$
    $x + y + 220° + 30° = 360°$ (angle sum of quadrilateral)
    $y + 307° = 360°$
    so $y = 53°$

    c) $c + c + c + 45° = 360°$ (angle sum of quadrilateral)
    $3c = 315°$
    so $c = 105°$

2. $85° + 85° + 85° +$ fourth angle $= 360°$ (angle sum of quadrilateral)
    $255° +$ fourth angle $= 360°$
    so fourth angle $= 105°$

3. a) Angle sum of an 11-sided polygon
    $= 11 \times 180° - 360°$
    $= 1980° - 360°$
    $= 1620°$

    b) Angle sum of a 32-sided polygon
    $= 32 \times 180° - 360°$
    $= 5760° - 360°$
    $= 5400°$

4. a) Angle sum of pentagon (5 sides) $= 5 \times 180° - 360°$
    $= 900° - 360°$
    $= 540°$

    Each interior angle of regular pentagon
    $= 540° \div 5 = 108°$.

    Alternative method
    The exterior angles add up to $360°$
    Each exterior angle $= 360° \div 5 = 72°$
    Each interior angle $= 180° - 72° = 108°$

4. b) Angle sum of a 15-sided polygon
    $= 15 \times 180° - 360°$
    $= 2700° - 360°$
    $= 2340°$

    Each interior angle of a regular 15-sided polygon
    $= 2340° \div 15$
    $= 156°$

    Alternative method
    The exterior angles add up to $360°$
    Each exterior angle $= 360° \div 15$
    $= 24°$
    Each interior angle $= 180° - 24°$
    $= 156°$

5. Angle sum of a 7-sided polygon
    $= 7 \times 180° - 360°$
    $= 1260° - 360°$
    $= 900°$

    Each interior angle of a regular 7-sided polygon
    $= 900° \div 7$

    a) As a mixed fraction, each interior angle $= 128\frac{4}{7}°$
    b) To the nearest degree, each interior angle $= 129°$

6. Angle sum of a hexagon (6 sides)
    $= 6 \times 180° - 360°$
    $= 1080° - 360°$
    $= 720°$

    Let angle BCD $= x$
    $100° + 100° + x + x + x + x = 720°$ (angle sum of hexagon)
    $200° + 4x = 720°$
    $4x = 520°$
    so $x = 130°$
    so angle BCD $= 130°$

7. Each exterior angle $= 180° - 170° = 10°$
    The exterior angles add up to $360°$ so the number of angles $= 360° \div 10° = 36°$.
    Hence, the number of sides $= 36$.

EXERCISE 6

1. a) 1 b) 2 c) 1 d) 4
    e) 3 f) 0 g) 1 h) 0
    i) 5

2. a) 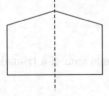 b)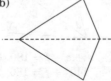

    c) 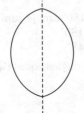 d)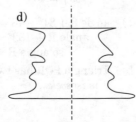

3. a) 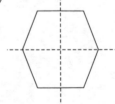 b)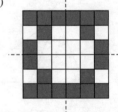

    c)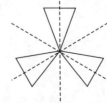

Module 4 Solutions

EXERCISE 6 (cont.)

4. a)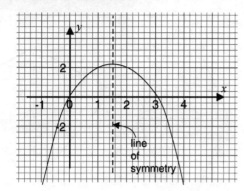

    b) The equation of the line of symmetry is $x = 1.5$.

5. a)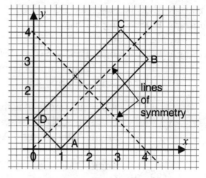

    b) The equations of the lines of symmetry are $y = x$ and $x + y = 4$.

EXERCISE 7

1. a) (i) 4   (ii) rotational symmetry of order 4
    b) (i) 2   (ii) rotational symmetry of order 2
    c) (i) 1   (ii) no rotational symmetry
    d) (i) 2   (ii) rotational symmetry of order 2
    e) (i) 3   (ii) rotational symmetry of order 3
    f) (i) 1   (ii) no rotational symmetry
    g) (i) 0   (ii) rotational symmetry of order 2
    h) (i) 0   (ii) rotational symmetry of order 6

2. a)

    b) The shape has no lines of symmetry.

3. a)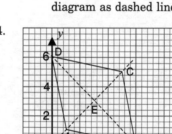

    b) The shape has 5 lines of symmetry (shown in the diagram as dashed lines).

4. The quadrilateral has 2 lines of symmetry (AC and BD). Their equations are $y = x$ and $x + y = 6$.

    The quadrilateral has rotational symmetry of order 2 about its centre E (where AC and BD cross).

EXERCISE 8

1. a) rhombus, square
    b) isosceles trapezium, rhombus, rectangle, square
    c) rectangle, square
    d) trapezium
    e) kite, rhombus, square
    f) trapezium, isosceles trapezium
    g) rhombus, square
    h) square
    i) trapezium, isosceles trapezium
    j) isosceles trapezium, kite

2. a)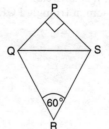

    A kite has two angles opposite each other which are equal. Angles P and R are not equal, so angles Q and S must be equal. Let them be $a°$.

    $$90° + a° + 60° + a° = 360° \text{ (angle sum of kite)}$$
    $$2a° + 150° = 360°$$
    $$2a° = 210°$$
    $$\text{so } a° = 105°$$
    so angles Q and S are each 105°

EXERCISE 8 (cont.)

2. b) Triangle PQS is isosceles (PS = PQ)
so angle PQS = angle PSQ
angle P = 90°
so angle PQS + angle PSQ + 90° = 180°
(angle sum of triangle PQS)
Hence, angle PQS = 45°

3.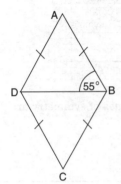
AB = AD (sides of a rhombus) so
triangle ABD is isosceles and
angle ADB = angle ABD = 55°

angle DAB + 55° + 55° = 180°
(angle sum of triangle ABD)

Hence, angle DAB = 70°
and so angle BCD = 70°
(opposite angle of the rhombus)

Also, angle DBC = 55° (diagonal
of rhombus bisects angles)
so angle ABC = 55° + 55° = 110°
The angles of rhombus ABCD are 70°, 110°, 70°, 110°.

4. a) b)
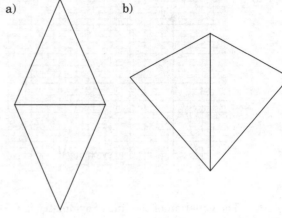
c)
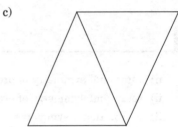

EXERCISE 9

1.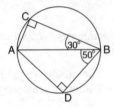

angle ACB = 90° (angle in a semicircle)
angle CAB + 90° + 30° = 180° (angle sum of
triangle ABC)
so angle CAB = 60°
angle ADB = 90° (angle in a semicircle)
angle BAD + 90° + 50° = 180° (angle sum of
triangle ABD)
so angle BAD = 40°
Hence, angle CAD = 60° + 40° = 100°

2. $p + 44° = 90°$ (angle between tangent and radius)
so $p = 46°$
and $q = 90°$ (angle in a semicircle)
$p + q + r = 180°$ (angle sum of triangle PQR)
$136° + r = 180°$
so $r = 44°$

3.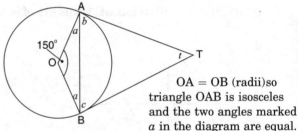

OA = OB (radii) so
triangle OAB is isosceles
and the two angles marked
a in the diagram are equal.

$a + a + 150° = 180°$ (angle sum of triangle OAB)
so $a = 15°$
$a + b = 90°$ (angle between tangent
and radius)
so $b = 75°$
$a + c = 90°$ (angle between tangent and radius)
so $c = 75°$
$b + c + t = 180°$ (angle sum of triangle ABT)
$75° + 75° + t = 180°$
so $t = 30°$

4.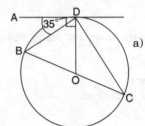

a) angle ADO = 90° (angle
between tangent and
radius)
so angle BDO = 90° − 35°
angle BDO = 55°

b) angle BDC = 90° (angle in a semicircle)
so angle ODC = 90° − angle BDO
= 90° − 55°
angle ODC = 35°

Check your progress 1

1. a) angle BDE + 115° = 180° (angles on a straight line)
    angle BDE = 65°
    b) angle ABD = angle BDE (alternate angles, AB parallel to CF)
    angle ABD = 65°
    c) angle BEF = 70° + angle BDE (exterior angle of triangle BDE)
    angle BEF = 135°

2. $2x + 116° = 180°$ (co-interior angles, AC parallel to ED)
    $2x = 64°$
    so $x = 32°$
    $x + y + 27° = 180°$ (angle sum of triangle ABC)
    $32° + y + 27° = 180°$
    $y + 59° = 180°$
    so $y = 121°$

3.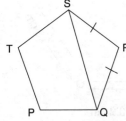

    a) Angle PQR is an interior angle of a regular pentagon. Angle sum of a pentagon
    $= 5 × 180° − 360°$
    $= 900° − 360°$
    $= 540°$
    Hence, angle PQR = 540° ÷ 5
    angle PQR = 108°

    b) In triangle QRS, angle QRS = 108° (angle of a regular pentagon)
    and QR = RS (sides of a regular pentagon)
    Hence, angle SQR + angle RSQ + 108° = 180° (angle sum of triangle QRS)
    and angle SQR = angle RSQ (base angles of isosceles triangle QRS)
    Hence, angle SQR = 72° ÷ 2 = 36°
    It follows that angle PQS = angle PQR − angle SQR = 108° − 36°
    angle PQS = 72°

4. a) Rotational symmetry of a regular hexagon is of order 6.
    b) Three possible solutions are shown below.

    c) The answer depends on which line you drew. The first two diagrams above each have 2 lines of symmetry (one 'horizontal' and one 'vertical'). The third diagram has no line of symmetry.

5. $a = 90°$ (angle between tangent and radius)
    $b + 37° = 90°$ (angle between tangent and radius)
    so $b = 53°$
    $c = 90°$ (angle in a semicircle)
    $c = 37° + d$ (exterior angle of triangle QST)
    $90° − 37° = d$
    so $d = 53°$

6. a) The star is not a regular polygon because its interior angles are not all equal.
    b) Angle sum of a 12-sided polygon
    $= 12 × 180° − 360°$
    $= 2160° − 360°$
    $= 1800°$
    c) The star shape can be regarded as two overlapping equilateral triangles.

    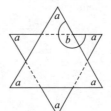

    Hence, $a = 60°$
    and $b = 180° + 60° = 240°$
    The star shape has six angles of 60° and six angles of 240°.

    > these 12 angles have a sum of 1800°, as expected by part (b)

    d) There are 6 lines of symmetry.
    e) Order of rotational symmetry is 6.

7. a) Each exterior angle = 180° − 175°.
    Each exterior angle = 5°.
    b) The exterior angles have a sum of 360° so the number of exterior angles = 360° ÷ 5° = 72°.
    Hence, the number of sides = 72.

EXERCISE 10

1. a) Triangle OAB is equilateral.
    b) Angle AOB = 60°.
    c) Each of the angles AOB, BOC, COD, DOE, EOF is 60°.
    Hence, the reflex angle AOF = 5 × 60° = 300°.
    Complete turn at O = 360° and so angle FOA = 60°.
    But OF = OA = 3 cm so triangle OFA is isosceles and has an angle of 60°.
    Hence, triangle OFA must be equilateral and FA = 3 cm.
    Thus the arc with centre F and radius 3 cm passes through A.

2. a) The sides are all equal in length.
    b) Taking the centre of the original circle to be O, angle ABO = 60° (angle of equilateral triangle OAB)
    and angle OBC = 60° (angle of equilateral triangle OBC)
    Hence angle ABC = 120°
    c) Each of the angles BCD, CDE, DEF, EFA and FAB is 120°.
    d) ABCDEF is a regular hexagon.

EXERCISE 10 (cont.)

3. a) Triangle OPR is an equilateral triangle.
    b) Angle POR = 60° because it is an angle of an equilateral triangle.

4. a) 2 cm
    b) You will find that this is a severe test of your accuracy!
    c) The arcs are less than half a circle so they are minor arcs.
    d) The diagram has 6 lines of symmetry.
    e) The diagram has rotational symmetry of order 6.

5. a) 1.5 cm
    b) You have to be very careful when drawing the semicircles.
    c) The diagram has no lines of symmetry.
    d) The diagram has rotational symmetry of order 2.

EXERCISE 11

Your answers may differ slightly from those shown below. Differences of 0.1 cm in lengths and 1° in angles are not unusual and would not be penalised by the IGCSE examiners.

1. a)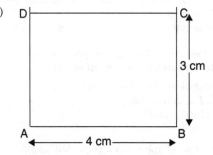

    b) AC = 5 cm and BD = 5 cm

2. a)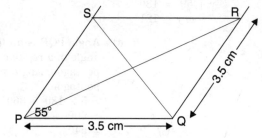

    b) PR = 6.2 cm and QS = 3.2 cm
    c) Angle between PR and QS = 90°

3. a)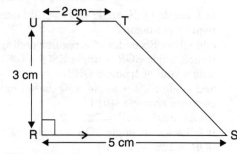

    b) ST = 4.25 cm
    c) Angle RST = 45°

EXERCISE 12

You are reminded that your answers may differ slightly from those shown below. If the difference is more than 0.1cm for a length or more than 1° for an angle, you should check your work carefully.

1. BC = 6.45 cm
    CA = 5.7 cm

    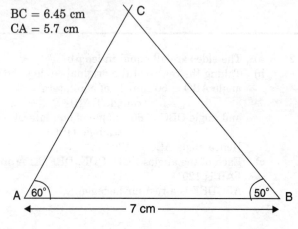

2. To draw the triangle, you need to calculate angle R.
    angle R = 180° − (70° + 30°)
    = 80°

    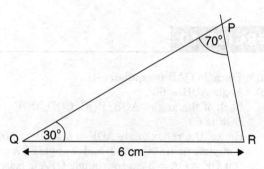

    Measuring the drawing,
    PQ = 6.3 cm and
    PR = 3.2 cm.

EXERCISE 12 (cont.)

3. angle Y = 69°
   angle Z = 51°
   side YZ = 5.6 cm

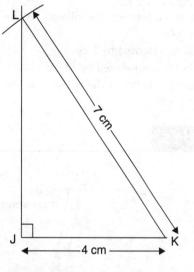

5. You should draw the side JK first and then the right angle at J. You need to use a compass to obtain the position of L – make sure that the circular arc is visible.

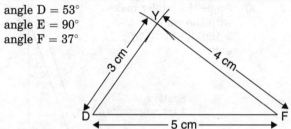

4. You must use a compass to draw this triangle. The circular arcs must be visible – do *not* rub them out!
   angle D = 53°
   angle E = 90°
   angle F = 37°

angle K = 55°
side JL = 5.75 cm

EXERCISE 13

1.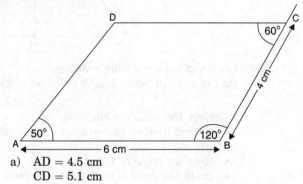

   a) AD = 4.5 cm
      CD = 5.1 cm
   b) ABCD is a *trapezium*. (AB is parallel to DC)

2. This can be done by drawing the sides and angles in this order:
   - side PQ = 4 cm
   - angle P = 120° and angle Q = 120°
   - side PU = 4 cm and side QR = 4 cm
   - angle U = 120° and angle R = 120°
   - side UT = 4 cm and side RS = 4 cm
   - join T to S to form side TS.

   However, an alternative and more accurate method is to draw a circle with radius 4 cm and follow the procedure in Exercise 10, Example 2.

2. (cont.)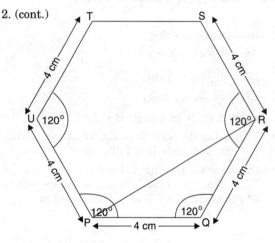

   Measuring the drawing, diagonal PR = 6.9 cm.

3.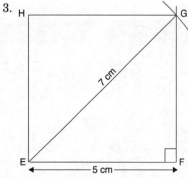

   The side EF should be drawn first and then the right angle at F. A compass must be used to locate the position of G; the circular arc must be visible. The position of H can be found by drawing right-angles at E and G or by drawing EH parallel to FG and GH parallel to FE.

   Measuring the drawing, FG = 4.9 cm.

EXERCISE 14

1. Length = 3.4 × 2 m = 6.8 m.
   Width = 2.6 × 2 m = 5.2 m.

2. a) 4 km is represented by 1 cm so 12 km is represented by 3 cm.
   The distance between the villages on the map = 3 cm.
   b) 5 km is represented by 1 cm
   so 12 km is represented by $\frac{12}{5}$ cm.
   The distance between the villages on the map = 2.4 cm.

3. a) 5 m is represented by 1 cm
   so 28 m is represented by $\frac{28}{5}$ cm.
   The length of the ramp on the drawing will be 5.6 cm.
   b) Angles in the drawing are the same as the angles in real life. On the drawing, the angle the ramp makes with the horizontal will be 15°.

EXERCISE 15

1.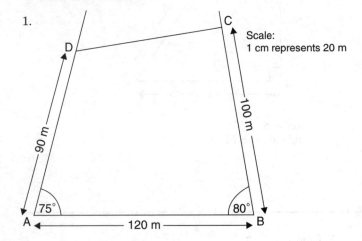

   a) In the scale drawing,
   AB = $\frac{120}{20}$ cm = 6 cm
   BC = $\frac{100}{20}$ cm = 5 cm
   AD = $\frac{90}{20}$ cm = 4.5 cm
   angle A = 75° and angle B = 80°.
   b) Measuring the scale drawing, we obtain angle C = 92° and angle D = 113°.
   c) On the scale drawing, CD = 4 cm.
   Using the scale (1 cm to 20 m), we deduce that the length of the side CD of the field = 80 m.

2. a) Angle the ladder makes with the wall
   = 180° − (90° + 70°)
   = 20°.

   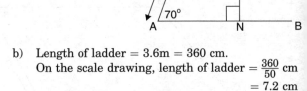

   b) Length of ladder = 3.6m = 360 cm.
   On the scale drawing, length of ladder = $\frac{360}{50}$ cm
   = 7.2 cm
   To obtain the scale drawing:
   • draw a line AB to represent the horizontal ground
   • draw an angle of 70° at A
   • mark the point C so that AC represents the ladder (AC = 7.2 cm)
   • draw an angle of 20° at C
   • mark the point N where the wall and the ground meet.

   On the scale drawing, CN = 6.75 cm.
   Using the scale (1 cm to 50 cm), we deduce that the distance the ladder reaches up the wall
   = 6.75 cm × 50 cm = 337.5 cm
   = 3.38 m (to 3 significant figures)

3. a) On the scale drawing, TF = 3.8 cm.
   Using the scale (1 cm represents 8 m), we deduce that the height of the building = 3.8 × 8 m
   = 30.4 m
   b) On the scale drawing, AF = 5.8 cm so the actual distance from A to F = 5.8 × 8 m = 46.4 m.
   c) Angle of elevation of the top of the building from A
   = angle TAF on the diagram
   = 34°

EXERCISE 16

1.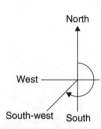

   Bearing for East = 090°

   Bearing for South-west = 180° + 45° = 225°

   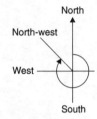 

   Bearing for North-west = 270° + 45° = 315°

   Bearing for North = 000°

2. a) Bearing of Johannesburg from Windhoek = 111°.
   b) Bearing of Johannesburg from Cape Town = 050°.
   c) Bearing of Cape Town from Johannesburg = 230°.
   d) Bearing of Lusaka from Cape Town = 028°.
   e) Bearing of Kimberley from Durban = 280°.

3.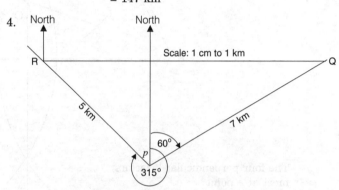

   a) Bearing of Bloemfontein from Kimberley = 108°.
   b) Bearing of Kimberley from Bloemfontein = 288°.
   c) On the scale drawing, the distance from Kimberley to Bloemfontein = 7.35 cm. Using the scale (1 cm to 20 km) the actual distance = 7.35 × 20 km = 147 km

4.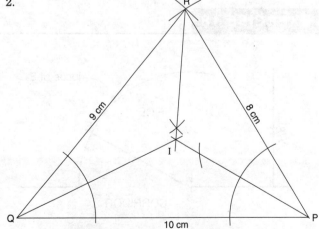

   a) Direct distance from village Q to village R = 9.6 km.
   b) Bearing of village Q from village R = 090°.

   > the question does not ask for a 3-figure bearing, so the answer 'East' is acceptable

EXERCISE 17

1.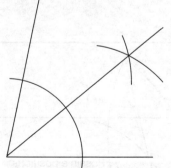

2. (diagram of triangle QPR with sides 9 cm, 8 cm, 10 cm and incentre I)

   The three angle bisectors meet at a point (marked I in the diagram).

   > this point is the centre of a circle which fits inside the triangle, touching all three sides

EXERCISE 17 (cont.)

3.

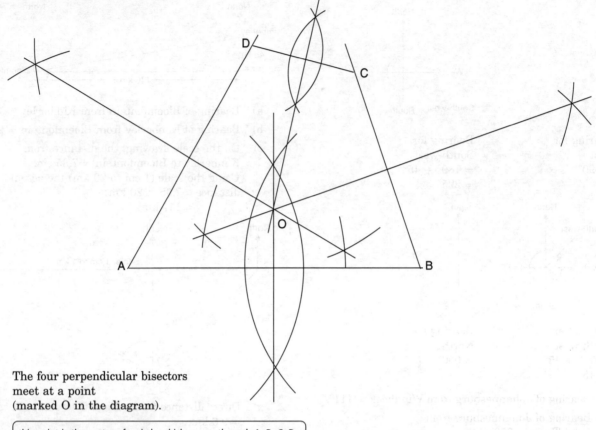

The four perpendicular bisectors
meet at a point
(marked O in the diagram).

> this point is the centre of a circle which passes through A, B, C, D

4. Draw an angle of 60° and bisect it.

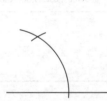 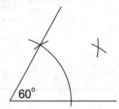 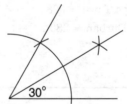

EXERCISE 18

1.

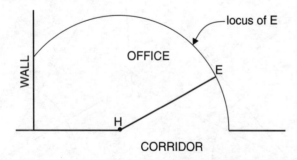

The locus of E is an arc of the circle which has centre H and radius equal to the length of HE.

2. In the diagram, (A) represents the locus of points equidistant from Lundu and Mish. (It is the perpendicular bisector of the line segment joining Lundu and Mish.)

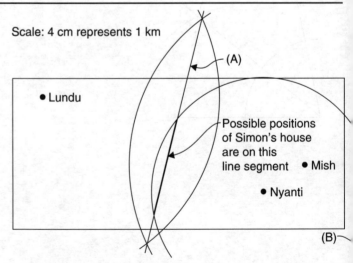

(B) represents the locus of points $\frac{3}{4}$ km from Nyanti. (It is a circle with centre at Nyanti and radius $\frac{3}{4}$ km.) Simon's house is on locus (A) and inside locus (B).

EXERCISE 18 (cont.)

3.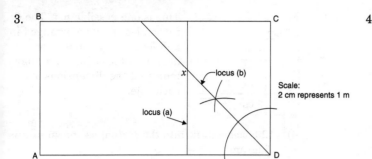

a) The locus is a line parallel to AB and 3 m from it.
b) The locus is the bisector of angle ADC.
c) (i) The point X is the intersection of locus (a) and locus (b).
   (ii) On the scale drawing, the distance of X from BC = 2 cm.
   The scale is 2 cm to 1 m, so the actual distance from X to BC = 1 m.

4.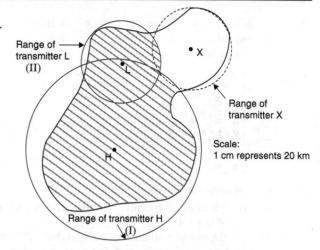

a) (I) is the locus of points 80 km from H.
   (II) is the locus of points 40 km from L.
   The part of the island in which television signals can be received is shaded in the diagram.
b) (i) A low powered transmitter will be sufficient.
   (ii) X marks a possible position for the transmitter.
   (The dashed circle indicates the range of the transmitter.)

Check your progress 2

1.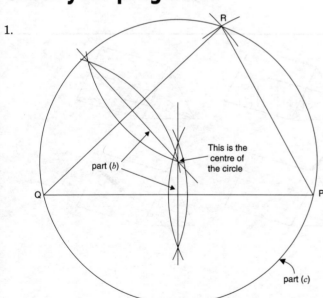

Notice that the construction arcs are clearly visible.

2.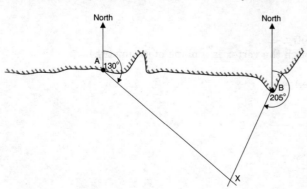

3. a) (i)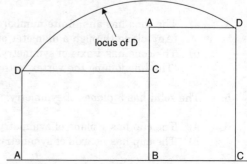

   (ii) The locus is the bisector of angle BCD.
   (iii) The locus is the circle with centre A and radius 6 m.
b) Points in the shaded region are nearer to CD than to CB and they are no more than 6 m from A.

4. The locus of D is an arc of a circle which has its centre at B.

EXERCISE 19

1. a) cube b) prism c) sphere
   d) cuboid e) cylinder f) cone
   g) pyramid

2.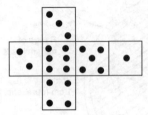

3. a) Pyramid (on a rectangular base)
   b) (i) PB must be 50 mm because it makes an edge of the solid with BQ.
       PA must be 50 mm because triangle PAB has two equal angles and so the triangle is isosceles.
      (ii) Triangle ADS will be equilateral.
      (iii) Angle SAD will be 60° (angle of an equilateral triangle).
   c) The solid has 5 faces, 8 edges and 5 vertices.

4. a) A cuboid (or rectangular block)
   b) 6 faces, 12 edges, 8 vertices
   c)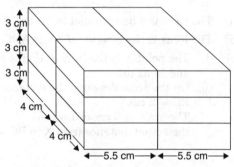
   This is one possible net. The 6 rectangles could be arranged in a different way. I have not drawn it to scale here but I have indicated the dimensions of each side.
   d) 12 boxes will fit into the carton, as shown in this diagram.

5. a) Number of cubes = $3 \times 3 \times 2 = 18$
   b) (i) 8 (the cubes at the corners of the cuboid)
      (ii) 8 (the cubes at the middle of the 4 edges at the top of the cuboid and the 4 edges at the bottom of the cuboid)
      (iii) 2 (the middle cubes of the top and bottom faces)

EXERCISE 20

1.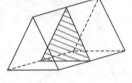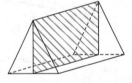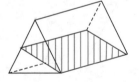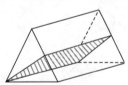

   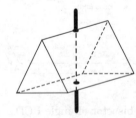 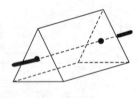 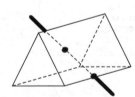 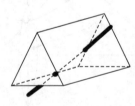

2. a) The cuboid has 5 planes of symmetry.
   b) The cuboid has 5 axes of symmetry.

3. a) The pyramid has 2 planes of symmetry.
   b) The pyramid has 1 axis of symmetry.

4. a) The cone has an infinite number of planes of symmetry.
      (Any plane through a diameter of the base and through the vertex is a plane of symmetry.)
   b) The cone has 1 axis of symmetry.
      (The line joining the vertex to the centre of the base.)

5. The solid has 5 planes of symmetry.

6. a) The cup has 1 plane of symmetry.
   b) The cup has no axis of symmetry.

Module 4 Solutions

EXERCISE 21

1. a) Figures D and F are congruent to figure A. (Note that F must be reflected to match A.)
   b) Figure H is congruent to figure E.

2. Figures Q, S and T are similar to figure P. (Note that S is *congruent* to P so they are the same shape and are similar to each other.)

3. a) Triangles AED and BEC are congruent.
      Triangles ACD and BDC are congruent.
      Triangles ABD and BAC are congruent.
      } Any one of these statements is an answer to the question.
   b) Triangles AEB and CED are similar but not congruent.

4.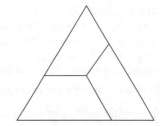

5. a) Ratio of the widths = $\frac{36 \text{ cm}}{12 \text{ cm}} = 3$
      Hence, ratio of the lengths = 3:1 and so the length of the poster = 3 × the length of the handbill. Therefore, the length of the handbill = $\frac{48}{3}$ cm = 16 cm.
   b) Ratio of the heights of the letters = 3:1 so the height of the letters on the poster = 9 mm × 3 = 27 mm (or 2.7 cm).

EXERCISE 22

1. a) Triangles JKL and QPR are similar (they have angles of 40°, 60°, 80°). Triangle OMD is not similar to the other two.
   b) Triangles STU and UVW are similar (their sides are in the ratio 1:2). Triangle XYZ is not similar to the other two.
   c) Triangles BCA and EFD are similar (they have angles of 20°, 30°, 130°). Triangle HIG is not similar to the other two.

2. a) The angles at the ends of the 5 cm side are 40° and 55° in both triangles. The triangles are congruent (ASA).
   b) The angles at the ends of the 7 cm side are 30° and 60° in both triangles. The triangles are congruent (ASA).
   c) The 3 sides of the first triangle are the same lengths as the sides of the second triangle. The triangles are congruent (SSS).
   d) The angle of 35° is not between the 3 cm and 4 cm sides in either triangle. There is insufficient information to determine whether the triangles are congruent or not.

3. a) In triangles RSX and PQX,
      angle R = angle P (alternate angles, SR parallel to PQ)
      angle S = angle Q (alternate angles, SR parallel to PQ)
      angle RXS = angle PXQ (vertically opposite angles)
      The 3 angles of triangle RSX are equal to the 3 angles of triangle PQX so the triangles are similar.
   b) $\frac{PQ}{RS} = \frac{QX}{SX} = \frac{XP}{XR}$ (ratio of corresponding sides)
      $\frac{7}{4} = \frac{QX}{3} = \frac{XP}{2}$
      Hence QX = $\frac{7 \times 3}{4}$ = 5.25 cm and
      XP = $\frac{7 \times 2}{4}$ = 3.5 cm

4.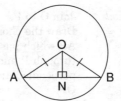
   In triangles OAN and OBN,
   OA = OB (radii)
   angle ONA = angle ONB = 90°
   ON = ON (same line in both triangles)
   The triangles OAN and OBN are congruent (RHS).

5.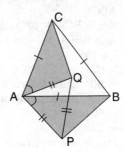
   a) angle CAQ = angle CAB − angle QAB
      = 60° − angle QAB (triangle ABC is equilateral)
      angle BAP = angle QAP − angle QAB
      = 60° − angle QAB (triangle QAP is equilateral)
      Hence, angle CAQ = angle BAP
   b) In triangles CAQ and BAP,
      angle CAQ = angle BAP (proved in part (a))
      CA = BA (sides of equilateral triangle ABC)
      AQ = AP (sides of equilateral triangle APQ)
      Hence, triangles CAQ and BAP are congruent (SAS).

EXERCISE 22 (cont.)

6. a) In triangles BDE and BAC,
angle BDE = angle BAC (corresponding angles, DE parallel to AC)
angle BED = angle BCA (corresponding angles, DE parallel to AC)
angle DBE = angle ABC (same angle)
Hence, the 3 angles of triangle BDE are equal to the 3 angles of triangle BAC, and so the triangles are similar.

b) $\frac{BD}{BA} = \frac{DE}{AC}$ (ratio of corresponding sides of similar triangles)
$\frac{BD}{2.4} = \frac{1.2}{1.8}$ and hence BD $= \frac{2.4 \times 1.2}{1.8} = 2.4 \times \frac{2}{3} = 1.6$ m

c) $\frac{BE}{BC} = \frac{DE}{AC}$ (ratio of corresponding sides of similar triangles)
$\frac{BE}{0.85 + BE} = \frac{1.2}{1.8}$ (BC = BE + EC = BE + 0.85)
$\frac{BE}{0.85 + BE} = \frac{2}{3}$
$3BE = 1.7 + 2BE$
$BE = 1.7$ m

EXERCISE 23

1.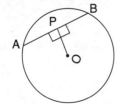

   Method   Join O to P. Draw the chord AB which passes through P and is perpendicular to OP.

   Explanation P is the midpoint of AB because OP is the perpendicular from the centre of the circle to the chord.

2. a) BA = BC (tangents from point B to the circle)
Hence, triangle ABC is isosceles.
   b) (i) angle CAB = angle BCA (base angles of isosceles triangle ABC)
and angle CAB + angle BCA + 40° = 180° (angle sum of triangle ABC)
Hence, angle CAB = 70°.
   (ii) angle DAC + angle CAB = 90° (angle between tangent and radius)
angle DAC + 70° = 90°
so angle DAC = 20°
   (iii) angle ACD = 90° (angle in a semicircle)
angle ADC + 20° + 90° = 180° (angle sum of triangle ACD)
so angle ADC = 70°

3.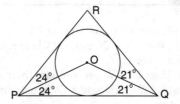

   angle OPQ = 48° ÷ 2 (OP bisects angle between tangents)
   = 24°
   angle OQP = 42° ÷ 2 (OQ bisects angle between tangents)
   = 21°
   angle POQ + 24° + 21° = 180° (angle sum of triangle OPQ)
   so angle OPQ = 135°

4. a) AB + BC + CA = (AR + RB) + (BP + PC) + (CQ + QA)
But QA = AR, RB = BP and CQ = PC (tangents from a point)
Hence, AB + BC + CA = 2(AR + BP + PC)
   b) AB = 14 cm, BC = 15 cm, CA = 13 cm and BP + PC = BC = 15 cm.
Substituting in the result of part (a):
14 + 15 + 13 = 2(AR + 15)
42 = 2(AR + 15)
21 = (AR + 15)
so AR = 6 cm
   c) BP = BR (tangents from a point)
But BR = AB − AR = 14 − 6 = 8 cm
Hence, BP = 8 cm

EXERCISE 24

1. a) $p = 2 \times 25°$ (angle at the centre theorem)
$p = 50°$
$p + q + r = 180°$ (angle sum of a triangle)
Hence, $q + r = 130°$
But $q = r$ (base angles of an isosceles triangle; equal radii)
so $q = 65°$ and $r = 65°$

   b) Note: The quadrilateral in the diagram is *not* cyclic; only 3 of its vertices are on the circumference of the circle.
reflex angle at O = 2 × 140° (angle at the centre theorem)
= 280°
$b + 280° = 360°$ (angles at a point)
so $b = 80°$

EXERCISE 24 (cont.)

1. c) $c = 30°$ (angles in the same segment)
        $d = 55°$ (angles in the same segment)
        $e + d = 100°$ (exterior angle of a triangle)
        so $e = 45°$
        $f = e$ (angles in the same segment)
        so $f = 45°$

2. a) $p + 95° = 180°$ (opposite angles of a cyclic quadrilateral)
        $p = 85°$
        $q + 115° = 180°$ (opposite angles of a cyclic quadrilateral)
        $q = 65°$

 b) $x = 102°$ (exterior angle of a cyclic quadrilateral = the opposite interior angle)
        $y + 96° = 180°$ (opposite angles of a cyclic quadrilateral)
        $y = 84°$

 c)   reflex angle O $= 360° - 130°$ (angles at a point)
                      $= 230°$
        $x = 230° \div 2$ (angle at the centre theorem)
        $x = 115°$
        $c = 180° - 115°$ (angles on a straight line)
        so $c = 65°$

3. a)   angle AOB $= 2x°$ (angle at the centre theorem)

 b)   angle OAB + angle OBA + $2x° = 180°$ (angle sum of triangle AOB)
        But OA = OB (radii)
        so angle OAB = angle OBA (base angles of isosceles triangle OAB)
        Hence, angle OAB $= (180° - 2x°) \div 2$
                 angle OAB $= 90° - x°$

 c)   angle BAT + angle OAB $= 90°$ (angle between tangent and radius)
        angle BAT $= 90° - (90° - x°)$
        so angle BAT $= x°$

Check your progress 3

1. a) (i) cuboid   (ii) prism   (iii) pyramid

 b)

Solid	Number of		
	Faces	Vertices	Edges
A	6	8	12
B	5	6	9
C	5	5	8
D	4	4	6
E	7	10	15
F	9	10	17

 c) number of faces + number of vertices = number of edges + 2

 > This is known as Euler's result for polyhedra. Leonard Euler was a Swiss mathematician who lived from 1707 to 1793.

2. a)   b)   c)

3. a) Triangles AZE, DXC, CDZ, EYD and BCY are congruent to triangle ABX.
 b) Triangles CXY, DYZ and ACD are similar to triangle ABX but not congruent to it.

4. a) The solid is a pyramid. It has 2 planes of symmetry.
 b) The solid has 1 axis of symmetry.

5. a) angle ABC $= 90°$ (angle in a semicircle)
      angle CAB $= 90° - x°$ (angle sum of triangle ABC)

 b) angle CAT $= 90°$ (angle between tangent and radius)
      angle BAT $= 90° - (90° - x°)$
      so angle BAT $= x°$
      Also angle ABT $= x°$ (TA = TB so triangle TAN is isosceles)
      angle ATB $+ x° + x° = 180°$ (angle sum of triangle TAB)
      angle ATB $= 180° - 2x°$

 c) angle ABP = angle CAB (alternate angles, BP is parallel to CA)
                    $= 90° - x°$
      angle PBT = angle ABT − angle ABP
                    $= x° - (90° - x°)$
      angle PBT $= 2x° - 90°$

Index

A
acute angle 4
acute-angled triangle 56
alternate angles 14
angle(s)
    acute 4
    alternate 14
    arms of 1
    calculating 10
    co-interior 14
    corresponding 13
    drawing 9
    exterior 18
    interior 18
    measuring 7
    obtuse 4
    of depression 64
    of elevation 64
    reflex 4
    right 2, 4
    straight 2, 4
arc
    major 48
    minor 48
    semicircular 48
axis of symmetry 96

B
bearings, 3-figure 68
bisect 37
bisector 72

C
calculating angles 10
centre (of a circle) 39
chord 39
circle(s) 39
    concentric 113
circumcircle (of a triangle) 84
circumference (of a circle) 39
co-interior 14
compass, using a 48
concentric circles 113
cone 90
congruent
    figures 99
    triangles 53, 106
convex polygon 18
corresponding angle 13

cube 88
cuboid 88
cyclic quadrilateral 115
cylinder 90

D
degree 2
depression, angle of 64
diagonal 23
diameter (of a circle) 39
drawing
    angles 9
    parallel lines 50
    perpendicular lines 51
    triangles 53

E
edge (of a solid) 88
elevation, angle of 64
equilateral triangle 20, 25
exterior angle 18

F
face (of a solid) 88
figures
    congruent 99
    similar 99
    solid 87

G
geometrical
    constructions 72
grad 3

I
interior angles 18
isosceles triangle 19

L
line(s)
    parallel 13
    symmetry 30
locus 74
    standard 76

M
major arc 48
measuring angles 7
minor arc 48

N
net 91

O
obtuse angle 4
obtuse-angled triangle 56
order of rotational
    symmetry 34

P
parallel lines 13
    drawing 50
perpendicular lines,
    drawing 50
plane of symmetry 95
polygon(s) 17, 24
    convex 18
    re-entrant 18
    regular 18, 25
polyhedron 88
prism, right 88
protractor 7
pyramid, right 89

Q
quadrilateral(s) 23
    cyclic 115

R
radian 3
radius 39
rectangular block 88
re-entrant polygon 18
reflex angle 4
regular polygon 18, 25
right
    angle 2, 4
    prism 88
    pyramid 89
right-angled triangle 56
rotational
    symmetry 34, 96
    order of 34

S
scale drawing 62
semicircle 39
semicircular arc 48
set square, using a 50

similar
    figures 99
    triangles 53, 103
sphere 90
square 25
straight angle 2, 4
symmetry 30
    axis of 96
    line 30
    of solids 95
    plane of 95
    rotational 34, 96

T
tangent 40
3-figure bearings 68
transversal 13
triangle(s) 19
    acute-angled 56
    congruent 53, 106
    drawing 53
    equilateral 20, 25
    isosceles 19
    obtuse-angled 56
    right-angled 56
    similar 53, 103

V
vertex 1, 88